Lecture Notes in Statistics

Edited by J. Berger, S. Fienberg, J. Gani,
K. Krickeberg, I. Olkin, and B. Singer

68

Masanobu Taniguchi

Higher Order Asymptotic Theory for Time Series Analysis

Springer-Verlag

Berlin Heidelberg New York London Paris
Tokyo Hong Kong Barcelona Budapest

Author

Masanobu Taniguchi
Department of Mathematical Science
Faculty of Engineering Science, Osaka University
Toyonaka 560, Japan

Mathematical Subject Classification: 62F12, 60G15, 62E20, 62F05, 62H10,
62M15

ISBN 0-387-97546-2 Springer-Verlag New York Berlin Heidelberg
ISBN 3-540-97546-2 Springer-Verlag Berlin Heidelberg New York

This work is subject to copyright. All rights are reserved, whether the whole or part of the material
is concerned, specifically the rights of translation, reprinting, re-use of illustrations, recitation,
broadcasting, reproduction on microfilms or in other ways, and storage in data banks. Duplication
of this publication or parts thereof is only permitted under the provisions of the German Copyright
Law of September 9, 1965, in its current version, and a copyright fee must always be paid.
Violations fall under the prosecution act of the German Copyright Law.

© Springer-Verlag Berlin Heidelberg 1991
Printed in Germany

Typesetting: Camera ready by author
Printing and binding: Druckhaus Beltz, Hemsbach/Bergstr.
2847/3140-543210 – Printed on acid-free paper

Cstack
osimon
1-9-92

To My Family

PREFACE

The initial basis of this book was a series of my research papers that I listed in References. I have many people to thank for the book's existence. Regarding higher order asymptotic efficiency I thank Professors Kei Takeuchi and M. Akahira for their many comments. I used their concept of efficiency for time series analysis.

During the summer of 1983, I had an opportunity to visit The Australian National University, and could elucidate the third-order asymptotics of some estimators. I express my sincere thanks to Professor E.J. Hannan for his warmest encouragement and kindness.

Multivariate time series analysis seems an important topic. In 1986 I visited Center for Multivariate Analysis, University of Pittsburgh. I received a lot of impact from multivariate analysis, and applied many multivariate methods to the higher order asymptotic theory of vector time series. I am very grateful to the late Professor P.R. Krishnaiah for his cooperation and kindness.

In Japan my research was mainly performed in Hiroshima University. There is a research group of statisticians who are interested in the *asymptotic expansions in statistics*. Throughout this book I often used the asymptotic expansion techniques. I thank all the members of this group, especially Professors Y. Fujikoshi and K. Maekawa for their helpful discussion.

When I was a student of Osaka University I learned multivariate analysis and time series analysis from Professors Masashi Okamoto and T. Nagai, respectively. It is a pleasure to thank them for giving me much of research background.

Finally, I am also indebted to Professor M. Huzii for my stay in Tokyo Institute of Technology in 1976.

Toyonaka, Japan
December 1990   Masanobu Taniguchi

Table of Contents

CHAPTER 1

A SURVEY OF THE FIRST-ORDER ASYMPTOTIC THEORY FOR TIME SERIES ANALYSIS

In this chapter we give a brief review of the first-order asymptotic theory for time series analysis and a motivation for the higher order asymptotic theory.

Whittle(1952, 1962) studied a method for estimating the parameters of a scalar-valued linear process $X(t) = \sum_{j=0}^{\infty} \alpha_j(\theta)e(t-j)$ with $\alpha_0(\theta) = 1$, where the $e(j)$ are i.i.d. random variables with mean zero and the innovation variance $\text{Var}\{e(j)\} = \sigma^2$ does not depend on θ. He proposed to estimate θ by a value $\hat{\theta}$ which minimizes $U(\theta) = \int_{-\pi}^{\pi} I_n(\lambda)/f_\theta(\lambda)\,d\lambda$, where $f_\theta(\lambda)$ is the spectral density of the process and $I_n(\lambda) = (2\pi n)^{-1}|\sum_{t=1}^{n} X(t)e^{i\lambda t}|^2$. For an estimator of σ^2 he proposed $\hat{\sigma}^2 = \int_{-\pi}^{\pi} I_n(\lambda)/g_{\hat{\theta}}(\lambda)\,d\lambda$, where $g_\theta(\lambda) = |\sum_{j=0}^{\infty} \alpha_j(\theta)e^{i\lambda j}|^2$. Walker(1964) and Hannan(1973) provided a rigorous asymptotic theory for the estimates $\hat{\theta}$ and $\hat{\sigma}^2$ under fairly general conditions. Since there are cases where the innovation variance σ^2 depends on θ, Hosoya(1974) and Dzhaparidze(1974) proposed to minimize

$$\int_{-\pi}^{\pi} \left\{ \log f_\theta(\lambda) + \frac{I_n(\lambda)}{f_\theta(\lambda)} \right\} d\lambda$$

instead of $U(\theta)$ in order to estimate θ, and they gave the asymptotic distribution of the estimator. For a vector-valued linear process Dunsmuir and Hannan(1976) and Dunsmuir(1979) investigated asymptotic properties of the estimator $\hat{\theta}$ which minimizes

$$\log \det K(\theta) + \int_{-\pi}^{\pi} \text{tr}\{f_\theta(\lambda)^{-1}I_n(\lambda)\}\,d\lambda$$

with respect to θ, where $K(\theta)$ is the innovation variance matrix and $f_\theta(\lambda)$ and $I_n(\lambda)$ are the spectral density and periodogram matrices, respectively. Furthermore Hosoya and Taniguchi(1982) developed an asymptotic theory for an estimator defined by the criterion

$$\int_{-\pi}^{\pi} \{\log \det f_\theta(\lambda) + \text{tr} f_\theta(\lambda)^{-1}I_n(\lambda)\}\,d\lambda,$$

under more natural and relaxed conditions. It is known that $-(n/4\pi)\int_{-\pi}^{\pi}\{\log f_\theta(\lambda)+I_n(\lambda)/f_\theta(\lambda)\}\,d\lambda$ and $-(n/2\sigma^2)\int_{-\pi}^{\pi} I_n(\lambda)/g_\theta(\lambda)\,d\lambda$ are, to within constant terms, approximations for the Gaussian likelihood and its quadratic form part, respectively (e.g., Dzhaparidze(1974)). To sum up, the above literatures showed the consistency and the asymptotic normality of "quasi-Gaussian maximum likelihood estimators". Here the asymptotic distribution is given by $N(0, I(\theta)^{-1})$, where $I(\theta)$ is the normalized limit of Fisher's information matrix. In this chapter we say that an estimator of θ is (first-order) asymptotically efficient if it has the asymptotic distribution $N(0, I(\theta)^{-1})$. The rigorous definition for asymptotic efficiency including higher-order case will be given in Chapter 2.

Now we give a wide class of asymptotically efficient estimators by a method which is essentially different from maximum likelihood estimation. Let $\{X_t; t = 0, \pm 1, \pm 2, \ldots\}$ be a scalar-valued linear process generated as

$$X_t = \sum_{j=0}^{\infty} G(j)e_{t-j},$$

where $E\{e_j\} = 0, E\{e_j^2\} = \sigma^2$ with $\sigma^2 > 0$, and $E\{e_j e_s\} = 0$ for $j \neq s$. We introduce $\mathbf{F}$, the space of spectral densities defined by

$$\mathbf{F} = \left\{ g; g(\lambda) = \frac{\sigma^2}{2\pi} \left| \sum_{j=0}^{\infty} G(j)e^{-ij\lambda} \right|^2, \right.$$

there exist $C < \infty$ and $\delta > 0$ such that

$$\left. \sum_{j=0}^{\infty}(1+j^2)|G(j)| \leq C, \left| \sum_{j=0}^{\infty} G(j)z^j \right| \geq \delta \text{ for all } |z| \leq 1 \right\}.$$

Assuming that $\{e_t\}$ is fourth order stationary, let $Q_e(t_1, t_2, t_3)$ be the joint fourth cumulant of $e_t, e_{t+t_1}, e_{t+t_2}, e_{t+t_3}$ and assume that

$$\sum_{t_1, t_2, t_3 = -\infty}^{\infty} |Q_e(t_1, t_2, t_3)| < \infty.$$

Then the process $\{e_t\}$ has a fourth-order spectral density $\tilde{Q}_e(\lambda_1, \lambda_2, \lambda_3)$ such that

$$\tilde{Q}_e(\lambda_1, \lambda_2, \lambda_3) = \left(\frac{1}{2\pi}\right)^3 \sum_{t_1, t_2, t_3 = -\infty}^{\infty} Q_e(t_1, t_2, t_3)e^{-i(\lambda_1 t_1 + \lambda_2 t_2 + \lambda_3 t_3)}.$$

Similarly we can define $Q_x(t_1, t_2, t_3)$ and $\tilde{Q}_x(\lambda_1, \lambda_2, \lambda_3)$ respectively, the fourth-order cumulant and spectral density of the process $\{X_t\}$. We propose to fit some parametric family $\mathbf{P} = \{f_\theta; f_\theta \in \mathbf{F}, \theta \in \Theta \subset R^P\}$ of spectral densities to the true spectral density $g(\lambda)$ by minimizing a minimum contrast type criterion. Initially, we make only the following assumption.

Assumption 1.1. $K(x)$ is a three times continuously differentiable function on $(0, \infty)$, and has a unique minimum at $x = 1$.

We then define the criterion which measures the nearness of f_θ to g by

$$D(f_\theta, g) = \int_{-\pi}^{\pi} K\{f_\theta(\lambda)/g(\lambda)\} \, d\lambda. \tag{1.1}$$

We give five examples of $D(f_\theta, g)$.

(1) $K(x) = \log x + 1/x$,

$$D(f_\theta, g) = \int_{-\pi}^{\pi} \{\log(f_\theta(\lambda)/g(\lambda)) + g(\lambda)/f_\theta(\lambda)\} \, d\lambda. \tag{1.2}$$

This criterion is equivalent to the quasi-Gaussian maximum likelihood type criterion $\int_{-\pi}^{\pi} \{\log f_\theta(\lambda) + g(\lambda)/f_\theta(\lambda)\} \, d\lambda$ (Taniguchi(1979), Hosoya and Taniguchi (1982)).

(2) $K(x) = -\log x + x$,

$$D(f_\theta, g) = \int_{-\pi}^{\pi} \{-\log(f_\theta(\lambda)/g(\lambda)) + f_\theta(\lambda)/g(\lambda)\}\, d\lambda. \tag{1.3}$$

(3) $K(x) = (\log x)^2$,

$$D(f_\theta, g) = \int_{-\pi}^{\pi} (\log f_\theta(\theta) - \log g(\lambda))^2\, d\lambda. \tag{1.4}$$

This is given in Taniguchi(1979) and (1981).

(4) $K(x) = x \log x - x$,

$$D(f_\theta, g) = \int_{-\pi}^{\pi} f_\theta(\lambda) g(\lambda)^{-1} \{\log(f_\theta(\lambda) g(\lambda)^{-1}) - 1\}\, d\lambda. \tag{1.5}$$

(5) $K(x) = (x^\alpha - 1)^2, \quad 0 < \alpha < \infty$

$$D(f_\theta, g) = \int_{-\pi}^{\pi} \{(f_\theta(\lambda)/g(\lambda))^\alpha - 1\}^2\, d\lambda. \tag{1.6}$$

Remark 1.1. For independently and identically distributed observations with true probability density $p(x)$, Beran(1977) discussed fitting a probability density model $q_\theta(x)$ by minimizing the Hellinger distance $\int |q_\theta(x)^{1/2} - p(x)^{1/2}|^2\, dx$. Also, Amari(1982) and Eguchi (1983) discussed fitting $q_\theta(x)$ by the criterion $D(p, q_\theta) = E_p[K\{q_\theta(x)/p(x)\}]$.

A functional T defined on **F** is determined by the requirement that for the parametric family of spectral densities **P**,

$$D(f_{T(g)}, g) = \min_{s \in \Theta} D(f_s, g), \text{ for every } g \in \mathbf{F}. \tag{1.7}$$

Since $T(g)$ may be multiple-valued, we shall use the notation $T(g)$ to indicate any one of the possible values, chosen arbitrarily. For $f \in \mathbf{F}$, denote the L_2-norm by $\|f\|^2 = \int_{-\pi}^{\pi} |f(\lambda)|^2\, d\lambda$. If, for $g_n, g \in \mathbf{F}, \|g_n - g\| \to 0$ as $n \to \infty$, then we denote $g_n \xrightarrow{L_2} g$. To ensure the existence of $T(g)$, some assumptions are needed on **P**.

Theorem 1.1. *Suppose that θ is a compact subset of R^p, that $\theta_1 \neq \theta_2$ implies $f_{\theta_1} \neq f_{\theta_2}$ on a set of positive Lebesgue measure, that $f_\theta(\lambda) \in \mathbf{P}$, and also that $f_\theta(\lambda)$ is continuous in θ. Then*
(a) *For every $g \in \mathbf{F}$, there exists a value $T(g) \in \Theta$ satisfying (1.7).*
(b) *If $T(g)$ is unique and if $g_n \xrightarrow{L_2} g$, then $T(g_n) \to T(g)$ as $n \to \infty$.*
(c) *$T(f_\theta) = \theta$ for every $\theta \in \Theta$.*

Proof. (a) Since $f_\theta, g \in \mathbf{F}, |f_\theta(\lambda)/g(\lambda)| \leq C^2/\delta^2$ for all $\lambda \in [-\pi, \pi]$. Thus, noting that $K(\cdot)$ satisfies Assumption 1.1, there exists $L > 0$ such that $|K\{f_\theta(\lambda)/g(\lambda)\}| \leq L$. Put $h(s) = D(f_s, g)$. Using the dominated convergence theorem, we have

$$|h(s_n) - h(s)| = \left| \int_{-\pi}^{\pi} \{K(f_{s_n}(\lambda)/g(\lambda)) - K(f_s(\lambda)/g(\lambda))\}\, d\lambda \right| \to 0,$$

for any sequence $\{ s_n \in \Theta, s_n \to s \}$. Hence h is continuous and achieves a minimum on the compact set Θ.
(b) Let $g_n \xrightarrow{L_2} g$, and put $h_n(s) = D(f_s, g_n)$. Noting that $K(\cdot)$ satisfies Assumption 1.1, we have

$$\lim_{n\to\infty} \sup_{s\in\Theta} |h_n(s) - h(s)|$$

$$= \lim_{n\to\infty} \sup_{s\in\Theta} \left| \int_{-\pi}^{\pi} [K\{f_s(\lambda)/g_n(\lambda)\} - K\{f_s(\lambda)/g(\lambda)\}] \, d\lambda \right|$$

$$= \lim_{n\to\infty} \sup_{s\in\Theta} \left| \int_{-\pi}^{\pi} \{g_n(\lambda) - g(\lambda)\} K'\{f_s(\lambda)/\tilde{g}_n(\lambda)\}\{-f_s(\lambda)/\tilde{g}_n(\lambda)^2\} \, d\lambda \right|, \qquad (1.8)$$

where $g_n(\lambda) \lesseqgtr \tilde{g}_n(\lambda) \lesseqgtr g(\lambda)$. Since $|f_s(\lambda)/\tilde{g}_n(\lambda)| \le C^2/\delta^2$ for all $\lambda \in [-\pi, \pi]$, Schwarz's inequality implies that (1.8) is dominated by $d_1\|g_n - g\|$ for some constant $d_1 > 0$. Thus, if $g_n \xrightarrow{L_2} g$,

$$\lim_{n\to\infty} \sup_{s\in\Theta} |h_n(s) - h(s)| = 0, \qquad (1.9)$$

which implies that $|\min_{s\in\Theta} h_n(s) - \min_{s\in\Theta} h(s)| \to 0$ i.e., $|h_n(T(g_n)) - h(T(g))| \to 0$. Also (1.9) implies that $|h_n(T(g_n)) - h(T(g_n))| \to 0$. Then we have $\lim_{n\to\infty} h(T(g_n)) = h(T(g))$. The uniqueness of $T(g)$ and the continuity of h imply $T(g_n) \to T(g)$.

(c) Remembering that $K(\cdot)$ satisfies Assumption 1.1, we have

$$\int_{-\pi}^{\pi} K\{f_s(\lambda)/f_\theta(\lambda)\} \, d\lambda \ge \int_{-\pi}^{\pi} K(1) \, d\lambda,$$

and the equality holds if and only if $f_s(\lambda) = f_\theta(\lambda)$ a.s., which implies the assertion.

We now impose a further assumption on f_θ.

Assumption 1.2. The spectral model $f_\theta(\lambda)$ is three times continuously differentiable with respect to θ, and every component of the second derivatives $\partial^2 f_\theta/\partial\theta\partial\theta'$ is continuous in λ.

Using Assumptions 1.1 and 1.2 we have,

Theorem 1.2. *Suppose that $T(g)$ exists uniquely and lies in $\mathrm{Int}(\theta)$, and that*

$$H_g = \int_{-\pi}^{\pi} \left[\frac{1}{g(\lambda)^2} K''\{f_\theta(\lambda)/g(\lambda)\} \frac{\partial}{\partial\theta} f_\theta(\lambda) \frac{\partial}{\partial\theta'} f_\theta(\lambda) + \frac{1}{g(\lambda)} K'\{f_\theta(\lambda)/g(\lambda)\} \frac{\partial^2}{\partial\theta\partial\theta'} f_\theta(\lambda) \right]_{\theta=T(g)} d\lambda$$

is a non-singular matrix. Then for every sequence of spectral densities $\{g_n\}$ satisfying $g_n \xrightarrow{L_2} g$, we have

$$T(g_n) = T(g) - \int_{-\pi}^{\pi} \rho_g(\lambda)\{g_n(\lambda) - g(\lambda)\} \, d\lambda + O\{\|g_n - g\|^2\}, \qquad (1.10)$$

where

$$\rho_g(\lambda) = -H_g^{-1} \left[K''\{f_\theta(\lambda)/g(\lambda)\} \frac{f_\theta(\lambda)}{g(\lambda)^3} + K'\{f_\theta(\lambda)/g(\lambda)\} \frac{1}{g(\lambda)^2} \right] \left[\frac{\partial}{\partial\theta} f_\theta(\lambda) \right]_{\theta=T(g)}.$$

Proof. By the definition of $T(g_n)$ and $T(g)$,

$$\int_{-\pi}^{\pi} K'\{f_\theta(\lambda)/g_n(\lambda)\} \frac{1}{g_n(\lambda)} \frac{\partial}{\partial\theta} f_\theta(\lambda) \Big|_{\theta=T(g_n)} d\lambda = 0, \qquad (1.11)$$

$$\int_{-\pi}^{\pi} K'\{f_\theta(\lambda)/g(\lambda)\}\frac{1}{g(\lambda)}\frac{\partial}{\partial\theta}f_\theta(\lambda)\bigg|_{\theta=T(g)}\,d\lambda = 0, \tag{1.12}$$

Then (1.11) can be written

$$\int_{-\pi}^{\pi}\Bigg[K'\{f_\theta(\lambda)/g_n(\lambda)\}\frac{1}{g_n(\lambda)}\frac{\partial}{\partial\theta}f_\theta(\lambda)\bigg|_{\theta=T(g)}$$
$$+\Bigg\{ K''(f_\theta(\lambda)/g_n(\lambda))\left(\frac{1}{g_n(\lambda)}\right)^2 \frac{\partial}{\partial\theta}f_\theta(\lambda)\frac{\partial}{\partial\theta'}f_\theta(\lambda)$$
$$+ K'(f_\theta(\lambda)/g_n(\lambda))\frac{1}{g_n(\lambda)}\frac{\partial^2}{\partial\theta\partial\theta'}f_\theta(\lambda)\Bigg\}_{\theta=\tilde\theta}(T(g_n)-T(g))\Bigg]\,d\lambda = 0,$$

where $\tilde\theta = T(g) + \tilde{A}(T(g_n)-T(g))$, $\tilde{A}$ being a $p\times p$-matrix.
In view of (1.12) we have

$$T(g_n)-T(g) = -\Bigg[\int_{-\pi}^{\pi}\Bigg\{ K''(f_\theta(\lambda)/g_n(\lambda))\frac{1}{g_n(\lambda)^2}\frac{\partial}{\partial\theta}f_\theta(\lambda)\frac{\partial}{\partial\theta'}f_\theta(\lambda)$$
$$+ K'(f_\theta(\lambda)/g_n(\lambda))\frac{1}{g_n(\lambda)}\frac{\partial^2}{\partial\theta\partial\theta'}f_\theta(\lambda)\Bigg\}_{\theta=\tilde\theta}\,d\lambda\Bigg]^{-1}$$
$$\times \int_{-\pi}^{\pi}\Bigg[K'\{f_\theta(\lambda)/g_n(\lambda)\}\frac{1}{g_n(\lambda)}\frac{\partial}{\partial\theta}f_\theta(\lambda) - K'\{f_\theta(\lambda)/g(\lambda)\}\cdot$$
$$\times \frac{1}{g(\lambda)}\frac{\partial}{\partial\theta}f_\theta(\lambda)\Bigg]_{\theta=T(g)}\,d\lambda. \tag{1.13}$$

Noting that $K(x)$ and $f_\theta(\lambda)$ are continuously three times differentiable with respect to x and θ respectively, and that $T(g_n)\to T(g)$ for $g_n \overset{L_2}{\to} g$, it is not difficult to show by Schwarz's inequality

$$\Bigg|\int_{-\pi}^{\pi}\Bigg[K''\{f_\theta(\lambda)/g_n(\lambda)\}\frac{1}{g_n(\lambda)^2}\frac{\partial}{\partial\theta}f_\theta(\lambda)\frac{\partial}{\partial\theta'}f_\theta(\lambda)$$
$$+ K'\{f_\theta(\lambda)/g_n(\lambda)\}\frac{1}{g_n(\lambda)}\frac{\partial^2}{\partial\theta\partial\theta'}f_\theta(\lambda)\Bigg]_{\theta=\tilde\theta}\,d\lambda - H_g\Bigg|$$
$$\leq d_2\|g_n - g\| + d_3|T(g_n)-T(g)|, \tag{1.14}$$

for some $d_2 > 0$ and $d_3 > 0$. Similarly we can show that

$$\int_{-\pi}^{\pi}\Bigg[K'\{f_\theta(\lambda)/g_n(\lambda)\}\frac{1}{g_n(\lambda)}\frac{\partial}{\partial\theta}f_\theta(\lambda) - K'\{f_\theta(\lambda)/g(\lambda)\}\frac{1}{g(\lambda)}\frac{\partial}{\partial\theta}f_\theta(\lambda)\Bigg]_{\theta=T(g)}\,d\lambda$$
$$= \int_{-\pi}^{\pi}\Bigg[K''\{f_\theta(\lambda)/g(\lambda)\}\left(-\frac{f_\theta(\lambda)}{g(\lambda)^3}\right) + K'\{f_\theta(\lambda)/g(\lambda)\}\left(\frac{-1}{g(\lambda)^2}\right)\frac{\partial}{\partial\theta}f_\theta(\lambda)\Bigg]_{\theta=T(g)}$$
$$\times \{g_n(\lambda)-g(\lambda)\}\,d\lambda + O\{\|g_n - g\|^2\}. \tag{1.15}$$

Then (1.13), (1.14) and (1.15) imply (1.10).

Corollary 1.2. *If $g = f_\theta$, then $T(g) = \theta$. Thus we have*

$$\rho_{f_\theta}(\lambda) = \left[\int_{-\pi}^{\pi}\frac{\partial}{\partial\theta}\log f_\theta(\lambda)\frac{\partial}{\partial\theta'}\log f_\theta(\lambda)\,d\lambda\right]^{-1}\frac{\partial}{\partial\theta}f_\theta^{-1}(\lambda).$$

Suppose that a stretch $(X_1, \ldots, X_n)$ of the series $\{X_t\}$ is available. We now construct estimators of $T(g)$. Since the true spectral density $g(\lambda)$ of $\{X_t\}$ is unknown, we estimate $T(g)$ by $T(\hat{g}_n)$, where $\hat{g}_n$ is a nonparametric window type estimator. We set down the following further assumptions.

Assumption 1.3. $W(x)$ is bounded, even, non-negative and such that

$$\int_{-\infty}^{\infty} W(x)\, dx = 1.$$

Assumption 1.4. For $M = O(n^\alpha)$, $(1/4 < \alpha < 1/2)$, the function $W_n(\lambda) = MW(M\lambda)$ can be expanded as $W_n(\lambda) = (2\pi)^{-1} \sum_l w(l/M)e^{-il\lambda}$, where $w(x)$ is a continuous, even function with $w(0) = 1$, $|w(x)| \leq 1$ and $\int_{-\infty}^{\infty} w(x)^2\, dx < \infty$, and satisfies

$$\lim_{|x| \to 0} \frac{1 - w(x)}{|x|^2} = \kappa_2 < \infty.$$

Henceforth we use the following nonparametric spectral estimator

$$\hat{g}_n(\lambda) = \int_{-\pi}^{\pi} W_n(\lambda - \mu) I_x(\mu)\, d\mu,$$

where $I_x(\mu) = (2\pi n)^{-1} |\sum_{t=1}^n X_t e^{it\mu}|^2$.

Since $g \in \mathbf{F}$, it is not difficult to check that the assumptions of Theorems 9 and 10 in Hannan(1970, Section V) are satisfied, whence

$$E\{\hat{g}_n(\lambda) - g(\lambda)\}^2 = O\left(\frac{M}{n}\right) + O(M^{-4}), \tag{1.16}$$

uniformly in λ (i.e., $\hat{g}_n(\lambda) - g(\lambda) = O_p\{(M/n)^{1/2}\}$).

Using Fubini's theorem for (1.16) we have,

Lemma 1.1. *Assume that $\hat{g}_n(\lambda)$ satisfies Assumptions 1.3 and 1.4. Then*

$$\|\hat{g}_n - g\| = O_p\{(M/n)^{1/2}\}. \tag{1.17}$$

Denote by $\mathbf{B}(t)$ the σ-field generated by $\{e_s : s \leq t\}$. To discuss the asymptotic theory, we impose the following conditions on the process $\{X_t\}$, as in Hosoya and Taniguchi(1982).

Assumption 1.5. (i) For each nonnegative integer s,

$$\text{Var}[E\{e_t e_{t+s} | \mathbf{B}(t - \tau)\} - \delta(s)\sigma^2] = O(\tau^{-2-\epsilon}), \epsilon > 0,$$

uniformly in t, where $\delta(s)$ is Kronecker's delta.

(ii) $E|E\{e_{t_1} e_{t_2} e_{t_3} e_{t_4} | \mathbf{B}(t_1 - \tau)\} - E\{e_{t_1} e_{t_2} e_{t_3} e_{t_4}\}| = O(\tau^{-1-\eta})$, uniformly in t_1, where $t_1 \leq t_2 \leq t_3 \leq t_4$ and $\eta > 0$.

Then, by Hosoya and Taniguchi(1982, p.150) we get,

Lemma 1.2. *Assume that the $\{X_t\}$ satisfies Assumption 1.5. Let $\psi(\lambda)$ be a $p \times 1$ vector-valued continuous function on $[-\pi, \pi]$ such that $\psi(\lambda) = \psi(-\lambda)$. Then*

$$J_n = \sqrt{n} \int_{-\pi}^{\pi} \psi(\lambda)[I_x(\lambda) - g(\lambda)] \, d\lambda$$

has, asymptotically, a normal distribution with zero mean vector and covariance matrix

$$V = 4\pi \int_{-\pi}^{\pi} \psi(\lambda)\psi(\lambda)'g(\lambda)^2 \, d\lambda + 2\pi \iint_{-\pi}^{\pi} \psi(\lambda_1)\psi(\lambda_2)'\tilde{Q}_x(-\lambda_1, \lambda_2, -\lambda_2) \, d\lambda_1 \, d\lambda_2.$$

Then we have the following theorem.

Theorem 1.3. *Suppose that Assumptions 1.3-1.5 are satisfied. For $\psi(\lambda)$ defined in Lemma 1.2, the distribution of the vector*

$$L_n = \sqrt{n} \int_{-\pi}^{\pi} \psi(\lambda)\{\hat{g}_n(\lambda) - g(\lambda)\} \, d\lambda$$

as $n \to \infty$ tends to the multivariate normal distribution $N(O_p, V)$.

Proof. In this proof, without loss of generality, we assume that $\psi(\lambda)$ is a scalar function. If we can show that $|J_n - L_n|$ converges to zero in probability, then the result follows from Lemma1.2. Now

$$
\begin{aligned}
L_n &= \sqrt{n} \int_{-\pi}^{\pi} \psi(\lambda) \left[\int_{-\pi}^{\pi} \{I_x(\mu) - g(\mu)\}W_n(\lambda - \mu) \, d\mu \right] d\lambda \\
&\quad + \sqrt{n} \int_{-\pi}^{\pi} \psi(\lambda) \left[\int_{-\pi}^{\pi} g(\mu)W_n(\lambda - \mu) \, d\mu - g(\lambda) \right] d\lambda \\
&= L_n^{(1)} + L_n^{(2)} \quad \text{(say).}
\end{aligned}
$$

Putting $M(\lambda - \mu) = \eta$, we have

$$L_n^{(1)} = \sqrt{n} \int_{-\pi}^{\pi} \int_{M(-\pi-\mu)}^{M(\pi-\mu)} \psi\left(\mu + \frac{\eta}{M}\right) W(\eta) \, d\eta \{I_x(\mu) - g(\mu)\} \, d\mu.$$

Then we can see that

$$|L_n^{(1)} - J_n| = \left| \sqrt{n} \int_{-\pi}^{\pi} A_M(\mu)\{I_x(\mu) - g(\mu)\} \, d\mu \right|,$$

where

$$A_M(\mu) = \int_{M(-\pi-\mu)}^{M(\pi-\mu)} \psi\left(\mu + \frac{\eta}{M}\right) W(\eta) \, d\eta - \psi(\mu) \int_{-\infty}^{\infty} W(\eta) \, d\eta.$$

By Lemma 1.2, we have

$$E|L_n^{(1)} - J_n|^2 = $$
$$4\pi \int_{-\pi}^{\pi} A_M(\mu)^2 g(\mu)^2 \, d\mu + 2\pi \iint_{-\pi}^{\pi} A_M(\mu_1)A_M(\mu_2)\tilde{Q}_x(-\mu_1, \mu_2, -\mu_2) \, d\mu_1 \, d\mu_2 + o(1). \quad (1.18)$$

For every $\epsilon > 0$, by the dominated convergence theorem we can show that

$$\lim_{M \to \infty} |A_M(\mu)| = 0 \quad \text{for} \quad \mu \in B_\epsilon,$$

where $B_\epsilon = [-\pi + \epsilon, \pi - \epsilon]$. Thus,

$$4\pi \int_{B_\epsilon} A_M(\mu)^2 g(\mu)^2 \, d\mu + 2\pi \iint_{B_\epsilon \times B_\epsilon} A_M(\mu_1) A_M(\mu_2) \tilde{Q}_x(-\mu_1, \mu_2, -\mu_2) \, d\mu_1 \, d\mu_2 \qquad (1.19)$$

converges to zero as $M \to \infty$. Since A_M, g and $\tilde{Q}_x$ are bounded in $B = [-\pi, \pi]$, there exists $d_4 > 0$ such that

$$\left| 4\pi \int_{B-B_\epsilon} A_M(\mu)^2 g(\mu)^2 \, d\mu \right.$$

$$\left. + 2\pi \iint_{B \times B - B_\epsilon \times B_\epsilon} A_M(\mu_1) A_M(\mu_1) \tilde{Q}_x(-\mu_1, \mu_2, -\mu_2) \, d\mu_1 \, d\mu_2 \right| \leq d_4 \epsilon. \qquad (1.20)$$

Since ϵ is chosen arbitrarily, (1.18), (1.19) and (1.20) imply $|L_n^{(1)} - J_n| \to 0$ in probability. Thus the proof is complete if we show $L_n^{(2)} \to 0$ as $n \to \infty$. Since $g \in \mathbf{F}$, it is easy to show that $\sum_{j=-\infty}^{\infty} |\gamma_x(j)| j^2 < \infty$ where $\gamma_x(j) = \mathrm{E}(X_t X_{t+j})$ (Brillinger(1975, p.78)). Using the bias evaluation method (e.g., Hannan(1970, p.283)), we can show that

$$\left| \int_{-\pi}^{\pi} g(\mu) W_n(\lambda - \mu) \, d\mu - g(\lambda) \right| = O(M^{-2}),$$

uniformly λ, which implies

$$L_n^{(2)} = O(\sqrt{n}/M^2) \to 0 \quad \text{as} \quad n \to \infty.$$

By Lemma 1.1 and Theorems 1.2 and 1.3 we have the following theorem.

Theorem 1.4. *Assume that Assumptions 1.1-1.5 hold. Suppose that $T(g)$ exists uniquely and lies in $\mathrm{Int}(\Theta)$, and that H_g is nonsingular. Then*

$$P - \lim_{n \to \infty} T(\hat{g}_n) = T(g),$$

and the limiting distribution of the vector $\sqrt{n}(T(\hat{g}_n) - T(g))$ under g, as $n \to \infty$, is multivariate normal with mean zero and covariance matrix

$$U = 4\pi \int_{-\pi}^{\pi} \rho_g(\lambda) \rho_g(\lambda)' g(\lambda)^2 \, d\lambda + 2\pi \iint_{-\pi}^{\pi} \rho_g(\lambda_1) \rho_g(\lambda_2)' \tilde{Q}_x(-\lambda_1, \lambda_2, -\lambda_2) \, d\lambda_1 \, d\lambda_2.$$

Remark 1.2. If $g = f_\theta$, then the above covariance matrix is equal to

$$\left[\int_{-\pi}^{\pi} \frac{\partial}{\partial \theta} \log f_\theta(\lambda) \frac{\partial}{\partial \theta'} \log f_\theta(\lambda) \, d\lambda \right]^{-1} \left[4\pi \int_{-\pi}^{\pi} \frac{\partial}{\partial \theta} \log f_\theta(\lambda) \frac{\partial}{\partial \theta'} \log f_\theta(\lambda) \, d\lambda \right.$$

$$\left. + 2\pi \iint_{-\pi}^{\pi} \frac{\partial}{\partial \theta} f_\theta(\lambda_1)^{-1} \frac{\partial}{\partial \theta'} f_\theta(\lambda_2)^{-1} \tilde{Q}_x(-\lambda_1, \lambda_2, -\lambda_2) \, d\lambda_1 \, d\lambda_2 \right] \left[\int_{-\pi}^{\pi} \frac{\partial}{\partial \theta} \log f_\theta(\lambda) \frac{\partial}{\partial \theta'} \log f_\theta(\lambda) \, d\lambda \right]^{-1}.$$

Here we add the following assumption.

Assumption 1.6. The fourth order cumulant of e_t satisfies

$$\mathrm{cum}\{e_{t_1}, e_{t_2}, e_{t_3}, e_{t_4}\} = \begin{cases} \kappa_4 & \text{if } t_1 = t_2 = t_3 = t_4, \\ 0 & \text{otherwise.} \end{cases}$$

Consider the estimation of an innovation-free parameter θ, i.e., one for which the relationship

$$\frac{\partial}{\partial\theta}\int_{-\pi}^{\pi} g(\lambda)/f_\theta(\lambda)\,d\lambda\bigg|_{\theta=T(g)} = 0$$

holds. Then it is easy to see that if θ is an innovation-free parameter, then

$$\iint_{-\pi}^{\pi} \frac{\partial}{\partial\theta} f_\theta(\lambda_1)^{-1}\frac{\partial}{\partial\theta'} f_\theta(\lambda_2)^{-1}\tilde{Q}_x(-\lambda_1,\lambda_2,-\lambda_2)\,d\lambda_1\,d\lambda_2 = 0$$

(see Hosoya and Taniguchi(1982, p.138)). If $g = f_\theta$, where θ is the innovation-free parameter, then, by Theorem 1.2, Corollary 1.2 and Theorem 1.3 we have

$$\sqrt{n}(T(\hat{g}_n) - \theta) = \sqrt{n}\int_{-\pi}^{\pi} \rho_{f_\theta}(\lambda)\{\hat{g}_n(\lambda) - g(\lambda)\}\,d\lambda + o_p(1),$$

which implies,

Theorem 1.5. *Suppose that Assumptions 1.1-1.6 holds. If $g = f_\theta$, where θ is the innovation-free parameter, then the limiting distribution of $\sqrt{n}(T(\hat{g}_n) - \theta)$ as $n \to \infty$, is multivariate normal with mean zero and covariance matrix*

$$I(\theta)^{-1} = 4\pi\left[\int_{-\pi}^{\pi} \frac{\partial}{\partial\theta}\log f_\theta(\lambda)\frac{\partial}{\partial\theta'}\log f_\theta(\lambda)\,d\lambda\right]^{-1}.$$

That is, any minimum contrast estimators $T(\hat{g}_n)$ are first-order asymptotically efficient under Assumptions 1.1-1.6.

Except for autoregressive models, (quasi) maximum likelihood estimation requires iterative computational procedures. Here we shall show that if we choose an appropriate $K(\cdot)$, our estimators $T(\hat{g}_n)$ give explicit, non-iterative and efficient estimators for various spectral parametrizations. Suppose that the spectral density $f_\theta(\lambda)$ is parametrized as

$$f_\theta(\lambda) = S\{A_\theta(\lambda)\}, \tag{1.21}$$

where $A_\theta(\lambda) = \sum_j \theta_j \exp(ij\lambda)$ and $S(\cdot)$ is a bijective continuously three times differentiable function. To give non-iterative estimators, the following relation should be imposed;

$$K\left[\frac{S\{A_\theta(\lambda)\}}{g(\lambda)}\right] = C_1(\lambda)A_\theta(\lambda)^2 + C_2(\lambda)A_\theta(\lambda) + C_3(\lambda) + C_4\log S\{A_\theta(\lambda)\}, \tag{1.22}$$

where $C_1(\lambda)$, $C_2(\lambda)$ and $C_3(\lambda)$ are functions which are independent of θ, and C_4 is a constant which is independent of θ and λ. Since $\int_{-\pi}^{\pi} C_4 \log S\{A_\theta(\lambda)\}\,d\lambda$ is a function of the innovation variance, if we estimate an innovation-free parameter $\theta = (\theta_1,\ldots,\theta_p)'$, then the derivative of $\int_{-\pi}^{\pi} C_4 \log S\{A_\theta(\lambda)\}\,d\lambda$ with respect to θ vanishes. For a spectral density $f_\theta(\lambda)$ given in (1.21), choose the function $K(\cdot)$ so that (1.22) and Assumption 1.1 are satisfied. Then, checking the above procedure we get the following results.

Theorem 1.6. *Suppose that Assumptions 1.1-1.6 hold. Let $T_{\natural}(\hat{g}_n)$ be the estimator which minimizes the criterion*

$$\int_{-\pi}^{\pi} K_{\natural}[f_\theta(\lambda)/\hat{g}_n(\lambda)]\,d\lambda$$

with respect to θ. We consider a $p \times p$-matrix $R_{\sharp} = [R_{\sharp}(j, l)]$, and a $p \times 1$-vector $\gamma_{\sharp} = [\gamma_{\sharp}(l)]$.

(i) In the case $f_\theta(\lambda) = (\sigma^2/2\pi)|\sum_{j=0}^{p} \theta_j e^{-ij\lambda}|^{-2}$, where $\theta_0 = 1$ and $\sum_{j=0}^{p} \theta_j z^j \neq 0$ for $|z| \leq 1$, choose $K_{AR}(x) = \log x + 1/x$, then the non-iterative estimator is given by $T_{AR}(\hat{g}_n) = -R_{AR}^{-1} \cdot \gamma_{AR}$, where

$$R_{AR}(j - l) = \int_{-\pi}^{\pi} \hat{g}_n(\lambda) \cos(j - l)\lambda \, d\lambda,$$

$$\gamma_{AR}(l) = \int_{-\pi}^{\pi} \hat{g}_n(\lambda) \cos l\lambda \, d\lambda.$$

(ii) In the case $f_\theta(\lambda) = (\sigma^2/2\pi)|\sum_{j=0}^{p} \theta_j e^{-ij\lambda}|^2$, where $\theta_0 = 1$ and $\sum_{j=0}^{p} \theta_j z^j \neq 0$ for $|z| \leq 1$, choose $K_{MA}(x) = -\log x + x$, then the non-iterative estimator is given by $T_{MA}(\hat{g}_n) = -R_{MA}^{-1} \cdot \gamma_{MA}$, where

$$R_{MA}(j - l) = \int_{-\pi}^{\pi} \hat{g}_n(\lambda)^{-1} \cos(j - l)\lambda \, d\lambda,$$

$$\gamma_{MA}(l) = \int_{-\pi}^{\pi} \hat{g}_n(\lambda)^{-1} \cos l\lambda \, d\lambda.$$

(iii) In the case $f_\theta(\lambda) = \sigma^2 \exp[\sum_{j=0}^{p} \theta_j \cos j\lambda]$, $\theta_0 = 1$ (cf. Bloomfield (1973)), choose $K_E(x) = (\log x)^2$, then the non-iterative estimator is given by

$$T_E(\hat{g}_n) = R_E^{-1} \cdot \gamma_E,$$

where

$$R_E(j - l) = \pi\delta(j - l) \quad and \quad \gamma_E(l) = \int_{-\pi}^{\pi} \cos l\lambda \cdot \log \hat{g}_n(\lambda) \, d\lambda.$$

(iv) Let $\Phi(x)$ be a three times continuously differentiable bijective function on $(0, \infty)$ and $\Phi(1) = 1$. We assume that there exists a function $h(\cdot)$ such that $\Phi(xy) = h(y)\Phi(x)$, for all $x, y \in (0, \infty)$. As a special case we can take $\Phi(x) = x^{1/\beta}, \beta > 0$.

In this case $f_\theta(\lambda) = \Phi^{-1}[\sum_{j=0}^{p} \theta_j \cos j\lambda]$, where $\theta_0 = 1$ and $\theta_1, \ldots, \theta_p$ satisfy $\sum_{j=0}^{p} \theta_j \cos j\lambda > 0$ for all $\lambda \in [-\pi, \pi]$, choose $K_\Phi(x) = [\Phi(x) - 1]^2$, then the non-iterative estimator is given by $T_\Phi(\hat{g}_n) = R_\Phi^{-1} \cdot \gamma_\Phi$, where

$$R_\Phi(j - l) = \int_{-\pi}^{\pi} h[\hat{g}_n(\lambda)^{-1}]^2 \cos j\lambda \cdot \cos l\lambda \, d\lambda,$$

$$\gamma_\Phi(l) = \int_{-\pi}^{\pi} h[\hat{g}_n(\lambda)^{-1}] \cos l\lambda \, d\lambda.$$

(v) If $g = f_\theta$, and if the process $\{X_t\}$ is Gaussian, then the above estimators $T_{AR}(\hat{g}_n)$, $T_{MA}(\hat{g}_n)$, $T_E(\hat{g}_n)$ and $T_\Phi(\hat{g}_n)$ are asymptotically efficient.

Remark 1.3. In the above (v) we assumed that $\{X_t\}$ is a Gaussian process. For the cases (i), (ii) and (iii), because the unknown parameters are innovation-free, the asymptotic covariance matrices of T_{AR}, T_{MA} and T_E attain the same first-order efficient bound $I(\theta)^{-1}$ even if $\{X_t\}$ is non-Gaussian. However, for the case (iv), since $(\theta_1, \ldots, \theta_p)$ is not always innovation-free, the asymptotic covariance matrix of T_Φ does not always attain the bound $I(\theta)^{-1}$ in the non-Gaussian case.

As we saw in the above, we could give a wide class of first-order asymptotically efficient estimators, which motivates higher order asymptotic theory in the following chapters.

CHAPTER 2

HIGHER ORDER ASYMPTOTIC THEORY FOR GAUSSIAN ARMA PROCESSES

2.1. Higher order asymptotic efficiency and Edgeworth expansions

In this chapter we give a survey of higher order asymptotic results in time series analysis, and explain a higher order asymptotic efficiency and a derivation of the Edgeworth expansion.

Recently some systematic studies of higher order asymptotic theory for stationary processes have been developed. For an autoregressive process of order 1 (AR(1)), Akahira(1975) showed that appropriately modified least squares estimator of the first-order coefficient θ is second-order asymptotically efficient in the class $\mathbf{A}_2$ of second-order asymptotically median unbiased (AMU) estimators if efficiency is measured by the degree of concentration of the sampling distribution up to second order. Also Akahira(1979) showed that the second-order asymptotic efficiency of a modified maximum likelihood estimator (MLE) of θ for AR(1) case.

Now, let $\{X_t; t = 0, \pm 1, \pm 2, \dots\}$ be a Gaussian autoregressive and moving average (ARMA) process with the spectral density $f_\theta(\lambda)$, $\theta \in R^1$, and mean 0. In this case Taniguchi(1983) showed that appropriately modified MLE of θ is second-order asymptotically efficient in the class $\mathbf{A}_2$. For an ARMA process with vector-valued unknown parameter $\boldsymbol{\theta}$ and mean $\mu \neq 0$, Tanaka(1984) gave the Edgeworth expansion of the joint distribution of MLE for $\boldsymbol{\theta}$ and μ up to second order.

Furthermore some results have appeared about the third-order asymptotic theory. Phillips(1977, 1978) derived the Edgeworth expansion for the distribution of the least squares estimator of the autoregressive coefficient in an AR(1) process up to third order. Ochi(1983) proposed a generalized estimator in the first-order autoregression, which includes the least squares estimator as a special case, and gave its third-order Edgeworth expansion. Also Fujikoshi and Ochi(1984) investigated the third-order asymptotic properties of the MLE and Ochi's generalized estimator. For ARMA processes, Taniguchi(1986) gave the third-order Edgeworth expansion for the MLE of a spectral parameter, and discussed its third-order asymptotic optimality in a certain class of estimators.

Throughout this book we use the following higher order asymptotic efficiency in the sense of highest probability concentration around the true value by the Edgeworth expansion. This concept of efficiency was introduced by Akahira and Takeuchi(1981). Let $\mathbf{X}_n = (X_1, \dots, X_n)'$ denote a sequence of random variables forming a stochastic process, and possessing the probability measure $P_\theta^n(\cdot)$, where $\theta \in \Theta$, a subset of the real line. If an estimator $\hat{\theta}_n$ of θ satisfies the equations

$$\lim_{n\to\infty} n^{(k-1)/2} \left| P_\theta^n \{ \sqrt{n}(\hat{\theta}_n - \theta) \leq 0 \} - \frac{1}{2} \right| = 0, \tag{2.1.1}$$

$$\lim_{n\to\infty} n^{(k-1)/2} \left| P_\theta^n \{ \sqrt{n}(\hat{\theta}_n - \theta) \geq 0 \} - \frac{1}{2} \right| = 0, \tag{2.1.2}$$

then $\hat{\theta}_n$ is called kth-order asymptotically median unbiased (kth-order AMU for short). We denote the set of kth-order AMU estimators by $\mathbf{A}_k$. For $\hat{\theta}_n$ kth-order AMU,

$$F_0^+(x,\theta) + n^{-1/2}F_1^+(x,\theta) + \cdots + n^{-(k-1)/2}F_{k-1}^+(x,\theta)$$

and

$$F_0^-(x,\theta) + n^{-1/2}F_1^-(x,\theta) + \cdots + n^{-(k-1)/2}F_{k-1}^-(x,\theta)$$

are said to be the kth-order asymptotic distributions of $\sqrt{n}(\hat{\theta}_n - \theta)$ if

$$
\begin{aligned}
\lim_{n\to\infty} n^{(k-1)/2} &\Big| P_\theta^n\{\sqrt{n}(\hat{\theta}_n - \theta) \le x\} - F_0^+(x,\theta) - n^{-1/2}F_1^+(x,\theta) \\
&- \cdots - n^{-(k-1)/2}F_{k-1}^+(x,\theta)\Big| = 0 \quad \text{for all} \quad x \ge 0,
\end{aligned}
\tag{2.1.3}
$$

$$
\begin{aligned}
\lim_{n\to\infty} n^{(k-1)/2} &\Big| P_\theta^n\{\sqrt{n}(\hat{\theta}_n - \theta) \le x\} - F_0^-(x,\theta) - n^{-1/2}F_1^-(x,\theta) \\
&- \cdots - n^{-(k-1)/2}F_{k-1}^-(x,\theta)\Big| = 0 \quad \text{for all} \quad x < 0.
\end{aligned}
\tag{2.1.4}
$$

For $\theta_0 \in \Theta$, consider the problem of testing hypothesis $H^+ : \theta = \theta_0 + x/\sqrt{n}$ $(x > 0)$ against alternative $K : \theta = \theta_0$.
We define

$$H_0^+(x,\theta_0) + n^{-1/2}H_1^+(x,\theta_0) + \cdots + n^{-(k-1)/2}H_{k-1}^+(x,\theta_0)$$

as follows

$$
\begin{aligned}
\sup_{\{A_n\}\in\Phi_X} \limsup_{n\to\infty} n^{(k-1)/2} &\Big\{ P_{\theta_0}^n(A_n) - H_0^+(x,\theta_0) \\
&- n^{-1/2}H_1^+(x,\theta_0) - \cdots - n^{-(k-1)/2}H_{k-1}^+(x,\theta_0)\Big\} = 0,
\end{aligned}
\tag{2.1.5}
$$

where Φ_X is the class of sets $A_n = \{\sqrt{n}(\tilde{\theta}_n - \theta) \le x\}$ with $\tilde{\theta}_n$ kth-order AMU.
Then we have for $x > 0$,

$$
\begin{aligned}
P_{\theta_0+x/\sqrt{n}}^n(A_n) &= P_{\theta_0+x/\sqrt{n}}^n\Big\{\sqrt{n}(\tilde{\theta}_n - \theta_0 - x/\sqrt{n}) \le 0\Big\} \\
&= \frac{1}{2} + o(n^{-(k-1)/2}).
\end{aligned}
\tag{2.1.6}
$$

By (2.1.3) and (2.1.5) we have

$$
\begin{aligned}
\limsup_{n\to\infty} n^{(k-1)/2} &\Big\{ F_0^+(x,\theta_0) + n^{-1/2}F_1^+(x,\theta_0) + \cdots \\
&+ n^{-(k-1)/2}F_{k-1}^+(x,\theta_0) - H_0^+(x,\theta_0) - n^{-1/2}H_1^+(x,\theta_0) \\
&- \cdots - n^{-(k-1)/2}H_{k-1}^+(x,\theta_0)\Big\} \le 0 \quad \text{for all} \quad x > 0.
\end{aligned}
\tag{2.1.7}
$$

Also consider the problem of the testing hypothesis $H^- : \theta = \theta_0 + x/\sqrt{n}$ $(x < 0)$ against alternative $K : \theta = \theta_0$.

Then we define

$$H_0^-(x,\theta_0) + n^{-1/2}H_1^-(x,\theta_0) + \cdots + n^{-(k-1)/2}H_{k-1}^-(x,\theta_0)$$

as follows

$$\inf_{\{A_n\}\in\Phi_X} \liminf_{n\to\infty} n^{(k-1)/2}\left\{ P_{\theta_0}^n(A_n) - H_0^-(x,\theta_0) \right.$$
$$\left. -n^{-1/2}H_1^-(x,\theta_0) - \cdots - n^{-(k-1)/2}H_{k-1}^-(x,\theta_0) \right\} = 0. \tag{2.1.8}$$

In the same way as for the case $x > 0$, by (2.1.4) and (2.1.8) we have

$$\liminf_{n\to\infty} n^{(k-1)/2}\left\{ F_0^-(x,\theta_0) + n^{-1/2}F_1^-(x,\theta_0) + \cdots \right.$$
$$+n^{-(k-1)/2}F_{k-1}^-(x,\theta_0) - H_0^-(x,\theta_0) - n^{-1/2}H_1^-(x,\theta_0)$$
$$\left. -\cdots - n^{-(k-1)/2}H_{k-1}^-(x,\theta_0) \right\} \geq 0 \quad \text{for all} \quad x < 0. \tag{2.1.9}$$

Thus we make the following definition.

Definition 2.1.1. (Akahira and Takeuchi(1981))
A kth-order AMU $\{\hat{\theta}_n\}$ is called kth-order asymptotically efficient if for each $\theta \in \Theta$,

$$\lim_{n\to\infty} P_\theta^n\left\{ \sqrt{n}(\hat{\theta}_n - \theta) \leq x \right\}$$
$$= H_0^+(x,\theta) + n^{-1/2}H_1^+(x,\theta) + \cdots + n^{-(k-1)/2}H_{k-1}^+(x,\theta) + o(n^{-(k-1)/2}) \quad \text{for all} \quad x \geq 0,$$
$$= H_0^-(x,\theta) + n^{-1/2}H_1^-(x,\theta) + \cdots + n^{-(k-1)/2}H_{k-1}^-(x,\theta) + o(n^{-(k-1)/2}) \quad \text{for all} \quad x < 0.$$

In the above discussion we can regard the bound distribution

$$H_0^+(x,\theta_0) + n^{-1/2}H_1^+(x,\theta_0) + \cdots + n^{-(k-1)/2}H_{k-1}^+(x,\theta_0)$$

as an approximation of the power function of the testing hypothesis $H^+ : \theta = \theta_0 + x/\sqrt{n}$ $(x > 0)$ against alternative $K : \theta = \theta_0$ at significance level $(1/2) + o(n^{-(k-1)/2})$.

By the fundamental lemma of Neyman and Pearson this bound distribution can be given by deriving the asymptotic expansion of the likelihood ratio test which tests the null hypothesis $H^+ :$ $\theta = \theta_0 + x/\sqrt{n}$ $(x > 0)$ against the alternative $K : \theta = \theta_0$ at significance level $(1/2) + o(n^{-(k-1)/2})$.
In case of $x < 0$, we can proceed similarly.

As we saw in the above we use the Edgeworth expansions of the concerned statistics in our discussions of higher order efficiency. Here we explain a derivation of the Edgeworth expansion.

Let $U_n = (u_1, \ldots, u_p)'$ be a measurable function of a sequence of random variables $X_1, \ldots, X_n$ forming a stochastic process. Suppose that all order of cumulants of U_n exist and satisfy the followings;

$$c_i = \text{cum}(u_i) = n^{-1/2}c_i^{(1)} + n^{-1}c_i^{(2)} + o(n^{-1}), \tag{2.1.10}$$

$$c_{ij} = \text{cum}(u_i, u_j) = c_{ij}^{(1)} + n^{-1/2}c_{ij}^{(2)} + n^{-1}c_{ij}^{(3)} + o(n^{-1}), \tag{2.1.11}$$

$$c_{ijk} = \text{cum}(u_i, u_j, u_k) = n^{-1/2}c_{ijk}^{(1)} + n^{-1}c_{ijk}^{(2)} + o(n^{-1}), \tag{2.1.12}$$

$$c_{ijkm} = \text{cum}(u_i, u_j, u_k, u_m) = n^{-1}c_{ijkm}^{(1)} + o(n^{-1}), \tag{2.1.13}$$

$i, j, k, m = 1, \ldots, p,$ and the Jth-order cumulant satisfies

$$c_{i_1 \ldots i_J} = \text{cum}^{(J)}(u_{i_1}, \ldots, u_{i_J}) = O(n^{-J/2+1}) \quad \text{for each} \quad J \geq 5.$$

Then the characteristic function of U_n is expressed as

$$\exp\left\{\sum_i c_i(it_i) + \frac{1}{2}\sum_{i,j} c_{ij}(it_i)(it_j) + \frac{1}{6}\sum_{i,j,k} c_{ijk}(it_i)(it_j)(it_k) + \frac{1}{24}\sum_{i,j,k,m} c_{ijkm}(it_i)(it_j)(it_k)(it_m) + \cdots\right\}$$

$$= \exp\left\{-\frac{1}{2}\sum_{i,j} c_{ij}^{(1)} t_i t_j\right\}\left[1 + \sum_i \left(\frac{c_i^{(1)}}{\sqrt{n}} + \frac{c_i^{(2)}}{n}\right)(it_i)\right.$$

$$+ \frac{1}{2}\sum_{i,j}\left(\frac{c_{ij}^{(2)}}{\sqrt{n}} + \frac{c_{ij}^{(3)}}{n} + \frac{c_i^{(1)}c_j^{(1)}}{n}\right)(it_i)(it_j)$$

$$+ \sum_{i,j,k}\left(\frac{c_{ijk}^{(1)}}{6\sqrt{n}} + \frac{c_{ijk}^{(2)}}{6n} + \frac{c_i^{(1)}c_{jk}^{(2)}}{2n}\right)(it_i)(it_j)(it_k)$$

$$+ \sum_{i,j,k,m}\left(\frac{c_{ijkm}^{(1)}}{24n} + \frac{c_{ij}^{(2)}c_{km}^{(2)}}{8n} + \frac{c_i^{(1)}c_{jkm}^{(1)}}{6n}\right)(it_i)(it_j)(it_k)(it_m)$$

$$+ \sum_{i,j,i',j',k'}\frac{c_{ij}^{(2)}c_{i'j'k'}^{(1)}}{12n}(it_i)(it_j)(it_{i'})(it_{j'})(it_{k'})$$

$$\left.+ \sum_{i,j,k,i',j',k'}\frac{c_{ijk}^{(1)}c_{i'j'k'}^{(1)}}{72n}(it_i)(it_j)(it_k)(it_{i'})(it_{j'})(it_{k'}) + o(n^{-1})\right]. \tag{2.1.14}$$

Inverting (2.1.14) by the Fourier inverse transform we have

$$P(u_1 < y_1, \ldots, u_p < y_p)$$

$$= \int_{-\infty}^{y_1} \cdots \int_{-\infty}^{y_p} N(\mathbf{y}; \Omega)\left[1 + \sum_i \left(\frac{c_i^{(1)}}{\sqrt{n}} + \frac{c_i^{(2)}}{n}\right)H_i(\mathbf{y})\right.$$

$$+ \frac{1}{2}\sum_{i,j}\left(\frac{c_{ij}^{(2)}}{\sqrt{n}} + \frac{c_{ij}^{(3)}}{n} + \frac{c_i^{(1)}c_j^{(1)}}{n}\right)H_{ij}(\mathbf{y})$$

$$+ \sum_{i,j,k}\left(\frac{c_{ijk}^{(1)}}{6\sqrt{n}} + \frac{c_{ijk}^{(2)}}{6n} + \frac{c_i^{(1)}c_{jk}^{(2)}}{2n}\right)H_{ijk}(\mathbf{y})$$

$$+ \sum_{i,j,k,m} \left(\frac{c^{(1)}_{ijkm}}{24n} + \frac{c^{(2)}_{ij} c^{(2)}_{km}}{8n} + \frac{c^{(1)}_{i} c^{(1)}_{jkm}}{6n} \right) H_{ijkm}(\mathbf{y})$$

$$+ \sum_{i,j,i',j',k'} \frac{c^{(2)}_{ij} c^{(1)}_{i'j'k'}}{12n} H_{iji'j'k'}(\mathbf{y})$$

$$+ \left. \sum_{i,j,k,i',j',k'} \frac{c^{(1)}_{ijk} c^{(1)}_{i'j'k'}}{72n} H_{ijki'j'k'}(\mathbf{y}) \right] d\mathbf{y} + o(n^{-1}) \qquad (2.1.15)$$

where $\mathbf{y} = (y_1, \ldots, y_p)'$,

$$N(\mathbf{y}; \Omega) = (2\pi)^{-p/2} |\Omega|^{-1/2} \exp \left(-\frac{1}{2} \mathbf{y}' \Omega^{-1} \mathbf{y} \right),$$

$$H_{j_1 \cdots j_s}(\mathbf{y}) = \frac{(-1)^s}{N(\mathbf{y}; \Omega)} \frac{\partial^s}{\partial y_{j_1} \cdots y_{j_s}} N(\mathbf{y}; \Omega),$$

and $\Omega = \{ c^{(1)}_{ij} \}$.

In the special case of $p = 1$ and $c^{(1)}_{11} = 1$, we have

$$P(u_1 < y_1)$$

$$= \Phi(y_1) - \phi(y_1) \left[\frac{c^{(1)}_{1}}{\sqrt{n}} + \frac{c^{(2)}_{1}}{n} + \frac{1}{2} \left(\frac{c^{(2)}_{11}}{\sqrt{n}} + \frac{c^{(3)}_{11}}{n} + \frac{c^{(1)}_{1} c^{(1)}_{1}}{n} \right) y_1 \right.$$

$$+ \left(\frac{c^{(1)}_{111}}{6\sqrt{n}} + \frac{c^{(2)}_{111}}{6n} + + \frac{c^{(1)}_{1} c^{(2)}_{11}}{2n} \right) (y_1^2 - 1)$$

$$+ \left(\frac{c^{(1)}_{1111}}{24n} + \frac{c^{(2)}_{11} c^{(2)}_{11}}{8n} + \frac{c^{(1)}_{1} c^{(1)}_{111}}{6n} \right) (y_1^3 - 3y_1)$$

$$+ \frac{c^{(2)}_{11} c^{(1)}_{111}}{12n} (y_1^4 - 6y_1^2 + 3)$$

$$+ \left. \frac{c^{(1)}_{111} c^{(1)}_{111}}{72n} (y_1^5 - 10y_1^3 + 15y_1) \right] + o(n^{-1}), \qquad (2.1.16)$$

where $\Phi(y) = \int_{-\infty}^{y} \phi(t)\, dt$, $\phi(t) = (1/\sqrt{2\pi}) \exp(-t^2/2)$.

These Edgeworth expansions are formal. As for the validity of Edgeworth expansions we will discuss it in Chapter 3.

2.2. Second-order asymptotic efficiency for Gaussian ARMA processes

In this section we investigate an optimal property of maximum likelihood and quasi-maximum likelihood estimators of Gaussian ARMA processes. It is shown that appropriate modifications of these estimators for Gaussian ARMA processes are second-order asymptotically efficient.

Now we present a basic theorem which enables us to evaluate the asymptotic cumulants (moments) of the maximum likelihood estimator. We introduce $\mathbf{D}_1$ and $\mathbf{D}_{ARMA}$, spaces of functions on $[-\pi, \pi]$ defined by

$$\mathbf{D}_1 = \left\{ f : f(\lambda) = \sum_{u=-\infty}^{\infty} a(u) \exp(-iu\lambda), a(u) = a(-u), \sum_{u=-\infty}^{\infty} |u||a(u)| < \infty \right\}, \quad (2.2.1)$$

$$\mathbf{D}_{ARMA} = \left\{ f : f(\lambda) = \frac{\sigma^2}{2\pi} \frac{\left| \sum_{j=0}^{q} a_j e^{ij\lambda} \right|^2}{\left| \sum_{j=0}^{p} b_j e^{ij\lambda} \right|^2}, (\sigma^2 > 0), \quad \text{for some positive} \right.$$

$$\text{integers } p \text{ and } q, \text{ where } A(z) = \sum_{j=0}^{q} a_j z^j \text{ and } B(z) =$$

$$\left. \sum_{j=0}^{p} b_j z^j \text{ are both bounded away from zero for } |z| \le 1 \right\} \qquad (2.2.2)$$

Noting Theorem 3.8.3 in Brillinger(1975), we have the following proposition.

Proposition 2.2.1.
(i) If $f_1, f_2 \in \mathbf{D}_1$, then $f_1 \cdot f_2 \in \mathbf{D}_1$.
(ii) If $f \in \mathbf{D}_{ARMA}$, then $f^{-1} \in \mathbf{D}_{ARMA}$.
(iii) If $f \in \mathbf{D}_{ARMA}$, then $f \in \mathbf{D}_1$.

For the subsequent discussions we introduce the following theorem.

Theorem 2.2.1. *Suppose that* $f_1(\lambda), \ldots, f_s(\lambda) \in \mathbf{D}_1, g_1(\lambda), \ldots, g_s(\lambda) \in \mathbf{D}_{ARMA}$. *We define* $\Gamma_1, \ldots, \Gamma_s, \Lambda_1, \ldots, \Lambda_s$, *the $n \times n$ Toeplitz type matrices, by*

$$\Gamma_j = \left(\int_{-\pi}^{\pi} e^{i(m_1 - m_2)\lambda} f_j(\lambda) \, d\lambda \right),$$

$$\Lambda_j = \left(\int_{-\pi}^{\pi} e^{i(m_1 - m_2)\lambda} g_j(\lambda) \, d\lambda \right),$$

$m_1, m_2 = 1, \ldots, n, j = 1, \ldots, s$. *If* $\phi^{(n)}(k), k = 1, \ldots, n$, *are the eigenvalues of* $\Gamma_1 \Lambda_1^{-1} \Gamma_2 \Lambda_2^{-1} \cdots \Gamma_s \Lambda_s^{-1}$, *then*

$$\frac{1}{n} \sum_{k=1}^{n} \phi^{(n)}(k) = \frac{1}{2\pi} \int_{-\pi}^{\pi} f_1(\lambda) \cdots f_s(\lambda) g_1(\lambda)^{-1} \cdots g_s(\lambda)^{-1} \, d\lambda + O(n^{-1}).$$

Proof. First, we show that each Λ_j is nonsingular. Since $g_j \in \mathbf{D}_{ARMA}$, there exist F_1, F_2 such that $0 < F_1 < g_j(\lambda) < F_2 < \infty$. If $\rho_1 \le \cdots \le \rho_n$ are the eigenvalues of Λ_j, we have $2\pi F_1 \le \rho_1 \le \cdots \le \rho_n \le 2\pi F_2$ (Grenander and Szegö(1958, p.64)), which implies the nonsingularity of Λ_j. Second we show that

$$n^{-1} \text{tr} \left\{ M_n(\psi_1) \cdots M_n(\psi_l) - M_n(\psi_1 \cdots \psi_l) \right\} = O(n^{-1}), \qquad (2.2.3)$$

where $\psi_1, \ldots, \psi_l \in \mathbf{D}_1$, and $M_n(\psi_j)$ is the $n \times n$-Toeplitz type matrix,

$$M_n(\psi_j) = \left(\frac{1}{2\pi} \int_{-\pi}^{\pi} e^{i(r-t)\lambda} \psi_j(\lambda) \, d\lambda \right), \quad r, t = 1, \ldots, n, \quad j = 1, \ldots, l.$$

Denote $m_{rt}(\psi_j)$ for the (r,t)-th element of $M_n(\psi_j)$. Since $\psi_j \in \mathbf{D}_1$, it follows that

$$\psi_j(\lambda) = \sum_{u=-\infty}^{\infty} \gamma_j(u)e^{-iu\lambda},$$

where

$$\sum_{u=-\infty}^{\infty} |u||\gamma_j(u)| < \infty, \quad j = 1, \ldots, l.$$

Let $S_n = (1/n)\mathrm{tr}\{M_n(\psi_1)\cdots M_n(\psi_l)\}$ and $L_n = (1/n)\mathrm{tr}\{M_n(\psi_1\cdots\psi_l)\}$. We have

$$
\begin{aligned}
S_n &= \frac{1}{n}\sum_{1\leq k_1,\ldots,k_l\leq n} m_{k_1 k_2}(\psi_1)m_{k_2 k_3}(\psi_2)\cdots m_{k_l k_1}(\psi_l) \\
&= \frac{1}{n}\sum_{1\leq k_1,\ldots,k_l\leq n} \frac{1}{2\pi}\int_{-\pi}^{\pi}\psi_1(\lambda)e^{i(k_1-k_2)\lambda}\,d\lambda \times \cdots \times \frac{1}{2\pi}\int_{-\pi}^{\pi}\psi_l(\lambda)e^{i(k_l-k_1)\lambda}\,d\lambda \\
&= \frac{1}{n}\sum_{1\leq k_1,\ldots,k_l\leq n} \gamma_1(k_1-k_2)\cdots\gamma_l(k_l-k_1) \\
&= \sum_{-n+1\leq j_1,\ldots,j_{l-1}\leq n-1} \gamma_1(j_1)\cdots\gamma_{l-1}(j_{l-1})\gamma_l(-j_1-\cdots-j_{l-1})\frac{n-K(j_1,\ldots,j_{l-1})}{n},
\end{aligned}
$$

where $K(j_1,\ldots,j_{l-1})$ is chosen suitably and satisfies

$$|K(j_1,\ldots,j_{l-1})| \leq |j_1| + \cdots + |j_{l-1}|.$$

On the other hand

$$
\begin{aligned}
L_n &= \frac{1}{2\pi}\int_{-\pi}^{\pi}\psi_1(\lambda)\cdots\psi_l(\lambda)\,d\lambda \\
&= \sum_{-\infty\leq j_1,\ldots,j_{l-1}\leq\infty} \gamma_1(j_1)\cdots\gamma_{l-1}(j_{l-1})\gamma_l(-j_1-\cdots-j_{l-1}).
\end{aligned}
$$

Thus we have

$$
\begin{aligned}
|S_n - L_n| &\leq {\sum}'|\gamma_1(j_1)\cdots\gamma_l(-j_1-\cdots-j_{l-1})| \\
&\quad + \sum_{|j_1|,\ldots,|j_{l-1}|\leq n-1} |\gamma_1(j_1)\cdots\gamma_l(-j_1-\cdots-j_{l-1})|\frac{|j_1|+\cdots+|j_{l-1}|}{n}, \quad (2.2.4)
\end{aligned}
$$

where $\sum' = \sum_{|j_1|,\ldots,|j_{l-1}|<\infty} - \sum_{|j_1|,\ldots,|j_{l-1}|\leq n-1}$.
The first term of the right hand side of (2.2.4) is bounded by

$$
\begin{aligned}
&\frac{1}{n}\sum_{k=1}^{l-1}\sum_{j_1=-\infty}^{\infty}\cdots\sum_{j_{k-1}=-\infty}^{\infty}\sum_{|j_k|\geq n}\sum_{j_{k+1}=-\infty}^{\infty}\cdots\sum_{j_{l-1}=-\infty}^{\infty} |j_k||\gamma_1(j_1)|\cdots|\gamma_l(-j_1-\cdots-j_{l-1})| \\
&\leq \frac{1}{n}\sum_{k=1}^{l-1}\sum_{j_1=-\infty}^{\infty}\cdots\sum_{j_{k-1}=-\infty}^{\infty}\sum_{|j_k|\geq n}\sum_{j_{k+1}=-\infty}^{\infty}\cdots\sum_{j_{l-1}=-\infty}^{\infty} |j_k||\gamma_1(j_1)|\cdots|\gamma_{l-1}(j_{l-1})|\sum_{j=-\infty}^{\infty}|\gamma_l(j)| \\
&= o(n^{-1}).
\end{aligned}
$$

The second term of the right hand side of (2.2.4) is bounded by

$$\frac{1}{n} \sum_{|j_1|,\ldots,|j_{l-1}|<\infty} (|j_1| + \cdots + |j_{l-1}|)\,|\gamma_1(j_1)| \cdots |\gamma_{l-1}(j_{l-1})| \sum_{j=-\infty}^{\infty} |\gamma_l(j)| = O(n^{-1}).$$

Thus we have completed the proof of (2.2.3). In the third step, we show that

$$\frac{1}{n}\mathrm{tr}\left\{ M_n(f_1)M_n(g_1)^{-1} \cdots M_n(f_{s-1})M_n(g_{s-1})^{-1}M_n(f_s)(M_n(g_s)^{-1} - M_n(g_s^{-1})) \right\} = O(n^{-1}). \quad (2.2.5)$$

Put

$$M = M_n(f_1)M_n(g_1)^{-1} \cdots M_n(f_{s-1})M_n(g_{s-1})^{-1}M_n(f_s)\left\{ M_n(g_s)^{-1} - M_n(g_s^{-1}) \right\}.$$

Then we have

$$\begin{aligned}
\frac{1}{n}\mathrm{tr}M &= \frac{1}{2n}\mathrm{tr}(M+M') \le \frac{1}{2n}\|M+M'\|\mathrm{rank}(M+M') \\
&\le \frac{1}{2n}(\|M\| + \|M'\|)(\mathrm{rank}M + \mathrm{rank}M') \\
&\le \frac{2}{n}\|M\|\mathrm{rank}M,
\end{aligned}$$

where $\|M\| = $ the square root of the largest eigenvalue of MM' (If M is symmetric, $\|M\| = $ the largest eigenvalue of M). Here we have

$$\begin{aligned}
\|M\| &\le \|M_n(f_1)\|\|M_n(g_1)^{-1}\| \cdots \|M_n(f_s)\|\|M_n(g_s)^{-1} - M_n(g_s^{-1})\| \\
&\le \|M_n(f_1)\|\|M_n(g_1)^{-1}\| \cdots \|M_n(f_s)\|\{\|M_n(g_s)^{-1}\| + \|M_n(g_s^{-1})\|\}.
\end{aligned}$$

Since there exist F_j and K_j such that $|f_j(\lambda)| \le F_j < \infty$ and $0 < K_j < g_j(\lambda)$, then

$$\|M_n(f_j)\| \le F_j, \quad \|M_n(g_j)^{-1}\| \le 1/K_j \quad \text{and} \quad \|M_n(g_s^{-1})\| \le 1/K_s$$

(Grenander and Szegö(1958, p.64)). Thus $\|M\|$ is bounded. Now

$$\begin{aligned}
\mathrm{rank}M &\le \mathrm{rank}\{M_n(g_s)^{-1} - M_n(g_s^{-1})\} \\
&= \min\{2\max(p,q),n\}
\end{aligned}$$

(Shaman(1976)) and this implies (2.2.5). Repeated use of (2.2.5) shows that

$$\frac{1}{n}\mathrm{tr}\{M_n(f_1)M_n(g_1)^{-1} \cdots M_n(f_s)M_n(g_s)^{-1} - M_n(f_1)M_n(g_1^{-1}) \cdots M_n(f_s)M_n(g_s^{-1})\} = O(n^{-1}).$$

By (2.2.3) we have

$$\frac{1}{n}\mathrm{tr}\{M_n(f_1)M_n(g_1)^{-1} \cdots M_n(f_s)M_n(g_s)^{-1} - M_n(f_1 g_1^{-1} \cdots f_s g_s^{-1})\} = O(n^{-1}),$$

which completes the proof.

Here we shall show that if we appropriately modify the Gaussian maximum likelihood estimator in an ARMA model, then it is second-order asymptotically efficient in the sense of Definition 2.2.1. In the first place we shall give the second-order bound distributions $H_0^+(x, \theta_0) + n^{-1/2}H_1^+(x, \theta_0)$ and $H_0^-(x, \theta_0) + n^{-1/2}H_1^-(x, \theta_0)$ defined by (2.1.5) and (2.1.8) respectively. Using the fundamental lemma of Neyman and Pearson these are given by the likelihood ratio test which tests the null hypothesis $H : \theta = \theta_0 + x/\sqrt{n}$ against the alternative $K : \theta = \theta_0$. In these discussions we use the formal Edgeworth expansions and stochastic expansions since their validities will be discussed in Chapter 3.

We now set down the following assumptions.

Assumption 2.2.1. The process $\{X_t; t = 0, \pm1, \pm2, \ldots\}$ is a Gaussian stationary process with the spectral density $f_\theta(\lambda) \in \mathbf{D}_{ARMA}, \theta \in \Theta \subset \mathbf{R}^1$, and mean 0.

Assumption 2.2.2. The spectral density $f_\theta(\lambda)$ is continuously three times differentiable with respect to θ, and the derivatives $\partial f_\theta/\partial\theta$, $\partial^2 f_\theta/\partial\theta^2$ and $\partial^3 f_\theta/\partial\theta^3$ belongs to $\mathbf{D}_1$.

Assumption 2.2.3. If $\theta_1 \neq \theta_2$, then $f_{\theta_1} \neq f_{\theta_2}$ on a set of positive Lebesgue measure.

Assumption 2.2.4.

$$I(\theta) = \frac{1}{4\pi}\int_{-\pi}^{\pi}\left\{\frac{\partial}{\partial\theta}\log f_\theta(\lambda)\right\}^2 d\lambda > 0, \quad \text{for all} \quad \theta \in \Theta.$$

Suppose that a stretch $\mathbf{X}_n = (X_1, \ldots, X_n)'$ of the series $\{X_t\}$ is available. Let Σ_n be the covariance matrix of $\mathbf{X}_n$. The (m, k)-th element of Σ_n is given by $\int_{-\pi}^{\pi}\exp\{i(m - k)\lambda\}f_\theta(\lambda)\,d\lambda$. The likelihood function based on $\mathbf{X}_n$ is given by

$$L(\theta) = (2\pi)^{-n/2}|\Sigma_n|^{-1/2}\exp\left(-\frac{1}{2}\mathbf{X}_n'\Sigma_n^{-1}\mathbf{X}_n\right).$$

Consider the problem of testing the hypothesis $H : \theta = \theta_0 + x/\sqrt{n}$ $(x > 0)$ against the alternative $K : \theta = \theta_0$. Let $LR = \log\{L(\theta_0)/L(\theta_1)\}$, where $\theta_1 = \theta_0 + x/\sqrt{n}$. If $\theta = \theta_0$, then we have

$$\begin{aligned}
LR &= -\frac{x}{\sqrt{n}}\left\{\frac{\partial}{\partial\theta}\log L(\theta)\right\}_{\theta_0} - \frac{x^2}{2n}\left\{\frac{\partial^2}{\partial\theta^2}\log L(\theta)\right\}\\
&\quad -\frac{x^3}{6n\sqrt{n}}\left\{\frac{\partial^3}{\partial\theta^3}\log L(\theta)\right\}_{\theta_0} + \text{ lower order terms.}
\end{aligned} \tag{2.2.6}$$

Now

$$\frac{\partial\log L(\theta)}{\partial\theta} = \frac{1}{2}\mathbf{X}_n'\Sigma_n^{-1}\dot{\Sigma}_n\Sigma_n^{-1}\mathbf{X}_n - \frac{1}{2}\text{tr}(\Sigma_n^{-1}\dot{\Sigma}_n), \tag{2.2.7}$$

$$\begin{aligned}
\frac{\partial^2\log L(\theta)}{\partial\theta^2} &= -\mathbf{X}_n'\Sigma_n^{-1}\dot{\Sigma}_n\Sigma_n^{-1}\dot{\Sigma}_n\Sigma_n^{-1}\mathbf{X}_n + \frac{1}{2}\mathbf{X}_n'\Sigma_n^{-1}\ddot{\Sigma}_n\Sigma_n^{-1}\mathbf{X}_n\\
&\quad -\frac{1}{2}\text{tr}(\Sigma_n^{-1}\ddot{\Sigma}_n - \Sigma_n^{-1}\dot{\Sigma}_n\Sigma_n^{-1}\dot{\Sigma}_n), \tag{2.2.8}
\end{aligned}$$

and

$$\frac{\partial^3 \log L(\theta)}{\partial \theta^3} = 3\mathbf{X}_n' \Sigma_n^{-1} \dot{\Sigma}_n \Sigma_n^{-1} \dot{\Sigma}_n \Sigma_n^{-1} \dot{\Sigma}_n \Sigma_n^{-1} \mathbf{X}_n - \frac{3}{2}\mathbf{X}_n' \Sigma_n^{-1} \ddot{\Sigma}_n \Sigma_n^{-1} \dot{\Sigma}_n \Sigma_n^{-1} \mathbf{X}_n$$

$$-\frac{3}{2}\mathbf{X}_n' \Sigma_n^{-1} \dot{\Sigma}_n \Sigma_n^{-1} \ddot{\Sigma}_n \Sigma_n^{-1} \mathbf{X}_n + \frac{1}{2}\mathbf{X}_n' \Sigma_n^{-1} \dddot{\Sigma}_n \Sigma_n^{-1} \mathbf{X}_n$$

$$-\frac{1}{2}\mathrm{tr}\{\Sigma_n^{-1} \dddot{\Sigma}_n - 3\Sigma_n^{-1}\dot{\Sigma}_n\Sigma_n^{-1}\ddot{\Sigma}_n + 2(\Sigma_n^{-1}\dot{\Sigma}_n)^3\}, \tag{2.2.9}$$

where $\dot{\Sigma}_n$, $\ddot{\Sigma}_n$ and $\dddot{\Sigma}_n$ are the $n \times n$ Toeplitz type matrices whose (m,k)-th elements are given by

$$\int_{-\pi}^{\pi} e^{i(m-k)\lambda} \frac{\partial}{\partial \theta} f_\theta(\lambda)\, d\lambda, \quad \int_{-\pi}^{\pi} e^{i(m-k)\lambda} \frac{\partial^2}{\partial \theta^2} f_\theta(\lambda)\, d\lambda, \quad \text{and} \quad \int_{-\pi}^{\pi} e^{i(m-k)\lambda} \frac{\partial^3}{\partial \theta^3} f_\theta(\lambda)\, d\lambda,$$

respectively. Since $\frac{\partial}{\partial \theta} f_\theta(\lambda)$, $\frac{\partial^2}{\partial \theta^2} f_\theta(\lambda)$ and $\frac{\partial^3}{\partial \theta^3} f_\theta(\lambda)$ belong to $\mathbf{D}_1$, using Theorem 2.2.1 we have

$$\frac{1}{n}E_{\theta_0}\left\{\frac{\partial^2}{\partial \theta^2}\log L(\theta)\right\}_{\theta_0} = -\frac{1}{2n}\mathrm{tr}\dot{\Sigma}_n\Sigma_n^{-1}\dot{\Sigma}_n\Sigma_n^{-1}$$

$$= -\frac{1}{4\pi}\int_{-\pi}^{\pi}\left\{\frac{\partial}{\partial\theta}f_\theta(\lambda)\right\}_{\theta_0}^2 f_{\theta_0}(\lambda)^{-2}\, d\lambda + O(n^{-1})$$

$$= -I(\theta_0) + O(n^{-1}), \tag{2.2.10}$$

and

$$\frac{1}{n}E_{\theta_0}\left\{\frac{\partial^3}{\partial \theta^3}\log L(\theta)\right\}_{\theta_0} = \frac{1}{n}\mathrm{tr}\{2(\dot{\Sigma}_n\Sigma_n^{-1})^3 - \frac{3}{2}\ddot{\Sigma}_n\Sigma_n^{-1}\dot{\Sigma}_n\Sigma_n^{-1}\}$$

$$= \frac{1}{\pi}\int_{-\pi}^{\pi}\left\{\frac{\partial}{\partial\theta}f_\theta(\lambda)\right\}_{\theta_0}^3 f_{\theta_0}(\lambda)^{-3}\, d\lambda$$

$$-\frac{3}{4\pi}\int_{-\pi}^{\pi}\left\{\frac{\partial^2}{\partial\theta^2}f_\theta(\lambda)\right\}_{\theta_0}\left\{\frac{\partial}{\partial\theta}f_\theta(\lambda)\right\}_{\theta_0} f_{\theta_0}(\lambda)^{-2}\, d\lambda + O(n^{-1})$$

$$= -3J(\theta_0) - K(\theta_0) + O(n^{-1}), \tag{2.2.11}$$

where

$$I(\theta) = \frac{1}{4\pi}\int_{-\pi}^{\pi}\left\{\frac{\partial}{\partial\theta}\log f_\theta(\lambda)\right\}^2 d\lambda,$$

$$J(\theta) = -\frac{1}{2\pi}\int_{-\pi}^{\pi}\left\{\frac{\partial}{\partial\theta}f_\theta(\lambda)\right\}^3 f_\theta(\lambda)^{-3}\, d\lambda + \frac{1}{4\pi}\int_{-\pi}^{\pi}\left\{\frac{\partial^2}{\partial\theta^2}f_\theta(\lambda)\right\}\left\{\frac{\partial}{\partial\theta}f_\theta(\lambda)\right\} f_\theta(\lambda)^{-2}\, d\lambda,$$

$$K(\theta) = \frac{1}{2\pi}\int_{-\pi}^{\pi}\left\{\frac{\partial}{\partial\theta}f_\theta(\lambda)\right\}^3 f_\theta(\lambda)^{-3}\, d\lambda.$$

Therefore, noting that $E_{\theta_0}\{\frac{\partial}{\partial\theta}\log L(\theta)\}_{\theta_0} = 0$, we have

$$E_{\theta_0}(LR) = \frac{x^2}{2}I(\theta_0) + \frac{x^3}{6\sqrt{n}}\{3J(\theta_0) + K(\theta_0)\} + O(n^{-1}). \tag{2.2.12}$$

To evaluate the higher order cumulants of LR, we need the following lemma, which is essentially due to Magnus and Neudecker(1979).

Lemma 2.2.1. *Let A, B and C be symmetric non-random matrices of order n. Then*

$$\text{(i)} \qquad \text{cum}\{\mathbf{X}'_n A \mathbf{X}_n, \mathbf{X}'_n B \mathbf{X}_n\} = 2\text{tr}A\Sigma_n B\Sigma_n, \qquad (2.2.13)$$

$$\text{(ii)} \qquad \text{cum}\{\mathbf{X}'_n A \mathbf{X}_n, \mathbf{X}'_n B \mathbf{X}_n, \mathbf{X}'_n C \mathbf{X}_n\} = 8\text{tr}A\Sigma_n B\Sigma_n C\Sigma_n. \qquad (2.2.14)$$

We can express LR such that

$$LR = \mathbf{X}'_n D_n \mathbf{X}_n + \text{non-random terms} + \text{stochastically lower order terms},$$

where

$$
\begin{aligned}
D_n &= -\frac{x}{2\sqrt{n}}\Sigma_n^{-1}\dot{\Sigma}_n\Sigma_n^{-1} + \frac{x^2}{2n}\Sigma_n^{-1}\dot{\Sigma}_n\Sigma_n^{-1}\dot{\Sigma}_n\Sigma_n^{-1} - \frac{x^2}{4n}\Sigma_n^{-1}\ddot{\Sigma}_n\Sigma_n^{-1} \\
&\quad -\frac{x^3}{2n\sqrt{n}}\Sigma_n^{-1}\dot{\Sigma}_n\Sigma_n^{-1}\dot{\Sigma}_n\Sigma_n^{-1}\dot{\Sigma}_n\Sigma_n^{-1} + \frac{x^3}{4n\sqrt{n}}\Sigma_n^{-1}\ddot{\Sigma}_n\Sigma_n^{-1}\dot{\Sigma}_n\Sigma_n^{-1} \\
&\quad +\frac{x^3}{4n\sqrt{n}}\Sigma_n^{-1}\dot{\Sigma}_n\Sigma_n^{-1}\ddot{\Sigma}_n\Sigma_n^{-1} - \frac{x^3}{12n\sqrt{n}}\Sigma_n^{-1}\,\dddot{\Sigma}_n\,\Sigma_n^{-1}.
\end{aligned}
$$

By (2.2.13) and Theorem 2.2.1 we have

$$
\begin{aligned}
\text{cum}_{\theta_0}\{LR, LR\} &= 2\text{tr}(D_n\Sigma_n)^2 + \text{lower order terms} \\
&= 2\text{tr}\left\{-\frac{x}{2\sqrt{n}}\Sigma_n^{-1}\dot{\Sigma}_n + \frac{x^2}{2n}(\Sigma_n^{-1}\dot{\Sigma}_n)^2 - \frac{x^2}{4n}\Sigma_n^{-1}\ddot{\Sigma}_n\right\}^2 + O(n^{-1}) \\
&= \frac{x^2}{2n}\text{tr}(\Sigma_n^{-1}\dot{\Sigma}_n)^2 - \frac{x^3}{n\sqrt{n}}\text{tr}(\Sigma_n^{-1}\dot{\Sigma}_n)^3 + \frac{x^3}{2n\sqrt{n}}\text{tr}\Sigma_n^{-1}\dot{\Sigma}_n\Sigma_n^{-1}\ddot{\Sigma}_n + O(n^{-1}) \\
&= x^2 I(\theta_0) + \frac{x^3}{\sqrt{n}}J(\theta_0) + O(n^{-1}). \qquad (2.2.15)
\end{aligned}
$$

By (2.2.14) and Theorem 2.2.1 we have

$$
\begin{aligned}
\text{cum}_{\theta_0}\{LR, LR, LR\} &= 8\text{tr}(D_n\Sigma_n)^3 + O(n^{-1}) \\
&= -8\text{tr}\frac{x^3}{8n\sqrt{n}}(\Sigma_n^{-1}\dot{\Sigma}_n)^3 + O(n^{-1}) \\
&= -\frac{x^3}{\sqrt{n}}K(\theta_0) + O(n^{-1}). \qquad (2.2.16)
\end{aligned}
$$

The Jth ($J \geq 3$) order cumulant of LR can be expressed as

$$
\begin{aligned}
\text{cum}_{\theta_0}^{(J)}(L) &= O\left[n^{-\frac{J}{2}}\text{cum}_{\theta_0}^{(J)}\{\mathbf{X}'_n\Sigma_n^{-1}\dot{\Sigma}_n\Sigma_n^{-1}\mathbf{X}_n\}\right] + \text{lower order terms} \\
&= O(n^{-\frac{J}{2}})\text{cum}_{\theta_0}^{(J)}\{\mathbf{U}'_n\Sigma_n^{-\frac{1}{2}}\dot{\Sigma}_n\Sigma_n^{-\frac{1}{2}}\mathbf{U}_n\} + \text{lower order terms}
\end{aligned}
$$

where $U'_n = (u_1, \ldots, u_n)$ is a random vector distributed as $N(O_n, I_n)$, (I_n is the $n \times n$-identity matrix). Denote the (i,j)-th component of $\Sigma_n^{-1/2} \dot{\Sigma}_n \Sigma_n^{-1/2}$ by a_{ij}. Then, using the fundamental properties of cumulant, we have

$$\operatorname{cum}_{\theta_0}^{(J)}(L) = O(n^{-\frac{J}{2}}) \sum_{i_1=1}^{n} \sum_{j_1=1}^{n} \cdots \sum_{i_J=1}^{n} \sum_{j_J=1}^{n} a_{i_1 j_1} \cdots a_{i_J j_J}$$
$$\times \operatorname{cum}_{\theta_0}^{(J)}\{u_{i_1} u_{j_1}, \ldots, u_{i_J} u_{j_J}\} + \text{ lower order terms},$$

(see Theorem 2.3.1 of Brillinger(1975)). Noting Theorem 2.3.2 of Brillinger(1975), it is not difficult to show

$$\operatorname{cum}_{\theta_0}^{(J)}(L) = O\{n^{-\frac{J}{2}} \cdot \operatorname{tr}(\dot{\Sigma}_n \Sigma_n^{-1})^J\}$$
$$= O(n^{-\frac{J}{2}+1}), \quad \text{(by Theorem 2.2.1)}. \qquad (2.2.17)$$

When $\theta = \theta_1$, in a similar way as the case $\theta = \theta_0$ we have

$$
\begin{aligned}
E_{\theta_1}(LR) = & -\frac{x}{\sqrt{n}} \operatorname{tr} \left\{ \frac{1}{2} \left(\frac{x}{\sqrt{n}} \dot{\Sigma}_n + \frac{x^2}{2n} \ddot{\Sigma}_n \right) \Sigma_n^{-1} \dot{\Sigma}_n \Sigma_n^{-1} \right\} \\
& -\frac{x^2}{2n} \left[\operatorname{tr} \left\{ -\left(\Sigma_n + \frac{x}{\sqrt{n}} \dot{\Sigma}_n \right) \Sigma_n^{-1} \dot{\Sigma}_n \Sigma_n^{-1} \dot{\Sigma}_n \Sigma_n^{-1} + \frac{1}{2} \left(\Sigma_n + \frac{x}{\sqrt{n}} \dot{\Sigma}_n \right) \Sigma_n^{-1} \ddot{\Sigma}_n \Sigma_n^{-1} \right\} \right. \\
& \left. -\frac{1}{2} \operatorname{tr} \{ \Sigma_n^{-1} \ddot{\Sigma}_n - (\Sigma_n^{-1} \dot{\Sigma}_n)^2 \} \right] - \frac{x^3}{6n\sqrt{n}} \operatorname{tr} \left\{ 3(\dot{\Sigma}_n \Sigma_n^{-1})^3 - \frac{3}{2} \ddot{\Sigma}_n \Sigma_n^{-1} \dot{\Sigma}_n \Sigma_n^{-1} \right. \\
& \left. -\frac{3}{2} \dot{\Sigma}_n \Sigma_n^{-1} \ddot{\Sigma}_n \Sigma_n^{-1} + \frac{1}{2} \dddot{\Sigma}_n \Sigma_n^{-1} \right\} \\
& -\frac{1}{2} \operatorname{tr} \{ \Sigma_n^{-1} \dddot{\Sigma}_n - 3\Sigma_n^{-1} \dot{\Sigma}_n \Sigma_n^{-1} \ddot{\Sigma}_n + 2(\Sigma_n^{-1} \dot{\Sigma}_n)^3 \} + O(n^{-1}) \\
= & -\frac{x^2}{2} I(\theta_0) - \frac{x^3}{6\sqrt{n}} \{ 3J(\theta_0) + 2K(\theta_0) \} + O(n^{-1}). \qquad (2.2.18)
\end{aligned}
$$

Similarly we can show that

$$\operatorname{cum}_{\theta_1}\{LR, LR\} = x^2 I(\theta_0) + \frac{x^3}{\sqrt{n}} \{J(\theta_0) + K(\theta_0)\} + O(n^{-1}), \qquad (2.2.19)$$

$$\operatorname{cum}_{\theta_1}\{LR, LR, LR\} = -\frac{x^3}{\sqrt{n}} K(\theta_0) + O(n^{-1}), \qquad (2.2.20)$$

$$\operatorname{cum}_{\theta_1}^{(J)}\{LR\} = O(n^{-\frac{J}{2}+1}), \quad \text{for} \quad J \geq 3. \qquad (2.2.21)$$

Applying the formula (2.1.15) to $LR + (x^2/2)I(\theta_0)$ under $\theta = \theta_1$, we obtain its Edgeworth expansion

$$P_{\theta_1}^n \left[LR + \frac{x^2}{2} I(\theta_0) \leq a \right]$$
$$= \Phi\left(\frac{a}{\sigma}\right) - \phi\left(\frac{a}{\sigma}\right) \left[\frac{c_1^{(1)}}{\sigma\sqrt{n}} + \frac{c_{11}^{(2)}}{2\sigma^2\sqrt{n}} \left(\frac{a}{\sigma}\right) + \frac{c_{111}^{(1)}}{6\sigma^3\sqrt{n}} \left\{\frac{a^2}{\sigma^2} - 1\right\} \right] + O(n^{-1}), \quad (2.2.22)$$

where $\sigma^2 = x^2 I(\theta_0)$, $c_1^{(1)} = -(x^3/6)\{3J(\theta_0) + 2K(\theta_0)\}$, $c_{11}^{(2)} = x^3\{J(\theta_0) + K(\theta_0)\}$ and $c_{111}^{(1)} = -x^3 K(\theta_0)$. Noting that $\Phi(x) = \Phi(0) + x\phi(0) + \cdots$, if we put $a = (1/\sqrt{n})c_1^{(1)} - (1/6\sigma^2\sqrt{n})c_{111}^{(1)}$ in (2.2.22), we can show that

$$P_{\theta_1}^n\left[LR + \frac{x^2}{2}I(\theta_0) \geq a\right] = \frac{1}{2} + O(n^{-1}). \tag{2.2.23}$$

Now putting $W_n = -\{LR + (x^2/2)I(\theta_0) - a - x^2 I(\theta_0)\}$, we have

$$E_{\theta_0}(W_n) = -\frac{x^3}{2\sqrt{n}}\{2J(\theta_0) + K(\theta_0)\} + \frac{x}{6\sqrt{n}}\frac{K(\theta_0)}{J(\theta_0)} + O(n^{-1}), \tag{2.2.24}$$

$$\text{cum}_{\theta_0}\{W_n, W_n\} = x^2 I(\theta_0) + \frac{x^3}{\sqrt{n}}J(\theta_0) + O(n^{-1}), \tag{2.2.25}$$

$$\text{cum}_{\theta_0}\{W_n, W_n, W_n\} = \frac{x^3}{\sqrt{n}}K(\theta_0) + O(n^{-1}), \tag{2.2.26}$$

$$\text{cum}_{\theta_0}^{(J)}(W_n) = O(n^{-\frac{J}{2}+1}), \quad \text{for} \quad J \geq 3. \tag{2.2.27}$$

By (2.2.24) – (2.2.27) and (2.1.15), we have

$$P_{\theta_0}^n[W_n \leq x^2 I(\theta_0)]$$
$$= \Phi\left\{x\sqrt{I(\theta_0)}\right\} + \frac{x^2}{6\sqrt{I(\theta_0)n}}\{3J(\theta_0) + 2K(\theta_0)\}\phi\left\{x\sqrt{I(\theta_0)}\right\} + O(n^{-1}). \tag{2.2.28}$$

If $\{\hat{\theta}_n\}$ is second-order AMU, then remembering (2.1.5) and the fundamental lemma of Neyman and Pearson, we have

Theorem 2.2.2.

$$\limsup_{n\to\infty} \sqrt{n}\left[P_{\theta_0}^n\left\{\sqrt{n}(\hat{\theta}_n - \theta_0) \leq x\right\}\right.$$
$$\left. -\Phi\left\{x\sqrt{I(\theta_0)}\right\} - \phi\left\{x\sqrt{I(\theta_0)}\right\}\left\{\frac{x^2}{6\sqrt{I(\theta_0)n}}(3J(\theta_0) + 2K(\theta_0))\right\}\right] \leq 0, \quad \text{for} \quad x \geq 0.$$

For $x < 0$, we have

$$\liminf_{n\to\infty} \sqrt{n}\left[P_{\theta_0}^n\left\{\sqrt{n}(\hat{\theta}_n - \theta_0) \leq x\right\}\right.$$
$$\left. -\Phi\left\{x\sqrt{I(\theta_0)}\right\} - \phi\left\{x\sqrt{I(\theta_0)}\right\}\left\{\frac{x^3}{6\sqrt{I(\theta_0)n}}(3J(\theta_0) + 2K(\theta_0))\right\}\right] \geq 0.$$

Remark 2.2.1. In the special case of

$$f_\theta(\lambda) = \frac{\sigma^2}{2\pi}\frac{1}{|1 - \theta e^{i\lambda}|^2},$$

i.e., an autoregressive model of order 1, the above bound distribution becomes

$$\Phi\left(x/\sqrt{1-\theta_0^2}\right) + \phi\left(x/\sqrt{1-\theta_0^2}\right)\left\{\theta_0 x^2(1-\theta_0^2)^{-\frac{3}{2}}\right\}/\sqrt{n},$$

which coincides with the results of Akahira(1975) and Akahira and Takeuchi(1981).

Now we define the maximum likelihood estimator $\hat{\theta}_{ML}$ of θ as a solution of the equation;

$$0 = \frac{\partial}{\partial\theta}\log L(\theta), \quad \text{for} \quad \theta \in \Theta. \tag{2.2.29}$$

By Theorem 2.2.1 and Lemma 2.2.1, it is easily shown that $\text{Var}[\frac{1}{n}\frac{\partial^3}{\partial\theta^3}\log L(\theta)] = O(n^{-1})$. Noting (2.2.11) we can show that

$$p - \lim_{n\to\infty} \frac{1}{n}\frac{\partial^3}{\partial\theta^3}\log L(\theta) = 3J(\theta) - K(\theta). \tag{2.2.30}$$

Expanding (2.2.29) as Taylor series, we obtain

$$\begin{aligned}
0 &= \frac{\partial}{\partial\theta}\log L(\hat{\theta}_{ML}) \\
&= \frac{\partial}{\partial\theta}\log L(\theta) + \left\{\frac{\partial^2}{\partial\theta^2}\log L(\theta)\right\}\left(\hat{\theta}_{ML} - \theta\right) + \frac{1}{2}\left\{\frac{\partial^3}{\partial\theta^3}\log L(\theta^*)\right\}\left(\hat{\theta}_{ML} - \theta\right)^2,
\end{aligned}$$

where $|\theta^* - \theta| \leq |\hat{\theta}_{ML} - \theta|$. Putting $U_n = \sqrt{n}(\hat{\theta}_{ML} - \theta)$, $Z_1(\theta) = \frac{1}{\sqrt{n}}\frac{\partial}{\partial\theta}\log L(\theta)$ and $Z_2(\theta) = \frac{1}{\sqrt{n}}\{\frac{\partial^2}{\partial\theta^2}\log L(\theta) - E_\theta(\frac{\partial^2}{\partial\theta^2}\log L(\theta))\}$, we have

$$0 = Z_1(\theta) + \frac{1}{\sqrt{n}}Z_2(\theta)U_n + \frac{1}{n}E_\theta\left(\frac{\partial^2}{\partial\theta^2}\log L(\theta)\right)U_n + \frac{1}{2n\sqrt{n}}\left\{\frac{\partial^3}{\partial\theta^3}\log L(\theta^*)\right\}U_n^2 \tag{2.2.31}$$

By (2.2.30) and (2.2.10), we can rewrite (2.2.31) as

$$0 = Z_1(\theta) + \left\{-I(\theta) + \frac{1}{\sqrt{n}}Z_2(\theta)\right\}U_n - \frac{1}{2\sqrt{n}}\{3J(\theta) + K(\theta)\}U_n^2 + o_p(n^{-\frac{1}{2}}),$$

which implies,

Theorem 2.2.3.

$$\sqrt{n}\left(\hat{\theta}_{ML} - \theta\right) = \frac{Z_1(\theta)}{I(\theta)} + \frac{Z_1(\theta)Z_2(\theta)}{I(\theta)^2\sqrt{n}} - \frac{3J(\theta) + K(\theta)}{2I(\theta)^3\sqrt{n}}Z_1(\theta)^2 + o_p(n^{-\frac{1}{2}}).$$

In the same way as the previous cumulant evaluations for LR, we can show that

$$E_\theta\{Z_1(\theta)^2\} = I(\theta) + O(n^{-1}), \tag{2.2.32}$$

$$E_\theta\{Z_1(\theta)Z_2(\theta)\} = J(\theta) + O(n^{-1}), \tag{2.2.33}$$

$$E_\theta\{Z_1(\theta)^3\} = \frac{1}{\sqrt{n}}K(\theta) + O(n^{-\frac{3}{2}}), \tag{2.2.34}$$

and the Jth $(J \geq 3)$ order cumulant of $Z_{j_1}(\theta), \ldots, Z_{j_J}(\theta)$ $(j_1, \ldots, j_J = 1$ or $2)$ satisfies

$$\text{cum}_\theta^{(J)}\{Z_{j_1}(\theta), \ldots, Z_{j_J}(\theta)\} = O(n^{-\frac{J}{2}+1}). \tag{2.2.35}$$

It follows from (2.2.32) – (2.2.35) and Theorem 2.2.3 that for $U_n = \sqrt{n}(\hat{\theta}_{ML} - \theta)$,

$$E_\theta U_n = -\frac{J(\theta) + K(\theta)}{2I(\theta)^2\sqrt{n}} + o(n^{-\frac{1}{2}}), \tag{2.2.36}$$

$$\mathrm{cum}_\theta\{U_n, U_n\} = \mathrm{cum}_\theta\left\{\frac{Z_1(\theta)}{I(\theta)}, \frac{Z_1(\theta)}{I(\theta)}\right\} + o(n^{-\frac{1}{2}})$$

$$= I(\theta)^{-1} + o(n^{-\frac{1}{2}}) \tag{2.2.37}$$

$$\mathrm{cum}_\theta\{U_n, U_n, U_n\} = \mathrm{cum}_\theta\left\{\frac{Z_1(\theta)}{I(\theta)}, \frac{Z_1(\theta)}{I(\theta)}, \frac{Z_1(\theta)}{I(\theta)}\right\}$$

$$+ \frac{3}{\sqrt{n}}\mathrm{cum}_\theta\left\{\frac{Z_1(\theta)}{I(\theta)}, \frac{Z_1(\theta)}{I(\theta)}, \frac{Z_1(\theta)Z_2(\theta)}{I(\theta)^2} - \frac{3J(\theta) + K(\theta)}{2I(\theta)^3}Z_1(\theta)^2\right\} + o(n^{-\frac{1}{2}})$$

$$= -\frac{3J(\theta) + 2K(\theta)}{I(\theta)^3\sqrt{n}} + o(n^{-\frac{1}{2}}) \tag{2.2.38}$$

$$\mathrm{cum}_\theta^{(J)}\{U_n\} = O(n^{-\frac{J}{2}+1}), \quad \text{for} \quad J \geq 4. \tag{2.2.39}$$

From (2.1.16) the Edgeworth expansion for $\hat{\theta}_{ML}$ is

$$P_\theta^n\left[\sqrt{I(\theta)n}(\hat{\theta}_{ML} - \theta) \leq y\right]$$

$$= \Phi(y) - \phi(y)\left\{-\frac{J(\theta) + K(\theta)}{2I(\theta)^{\frac{3}{2}}\sqrt{n}} - \frac{3J(\theta) + 2K(\theta)}{6I(\theta)^{\frac{3}{2}}\sqrt{n}}(y^2 - 1)\right\} + o(n^{-\frac{1}{2}}). \tag{2.2.40}$$

Putting $y = 0$ in (2.2.40) we have

$$P_\theta^n\left[\sqrt{I(\theta)n}(\hat{\theta}_{ML} - \theta) \leq 0\right] = \frac{1}{2} + \phi(0)\frac{K(\theta)}{6I(\theta)^{\frac{3}{2}}\sqrt{n}} + o(n^{-\frac{1}{2}}),$$

which implies that $\hat{\theta}_{ML}$ is not second-order AMU. If we put

$$\hat{\theta}_{ML}^* = \hat{\theta}_{ML} + \frac{K(\hat{\theta}_{ML})}{6nI(\hat{\theta}_{ML})^2} = \hat{\theta}_{ML} + \frac{K(\theta)}{6nI(\theta)^2} + o_p(n^{-1}),$$

then we obtain

$$P_\theta^n\left[\sqrt{n}(\hat{\theta}_{ML}^* - \theta) \leq 0\right] = \frac{1}{2} + o(n^{-\frac{1}{2}}), \tag{2.2.41}$$

($\hat{\theta}_{ML}^*$ is a second-order AMU), and

$$P_\theta^n\left[\sqrt{n}(\hat{\theta}_{ML}^* - \theta) \leq x\right]$$

$$= \Phi\left\{x\sqrt{I(\theta)}\right\} + \frac{x^2}{6\sqrt{I(\theta)n}}\{3J(\theta) + 2K(\theta)\}\phi\left\{x\sqrt{I(\theta)}\right\} + o(n^{-\frac{1}{2}}). \tag{2.2.42}$$

Remembering Theorem 2.2.2, we can see that (2.2.42) coincides with the bound distribution. Thus we have

Theorem 2.2.4. *The modified maximum likelihood estimator $\hat{\theta}_{ML}^*$ is second-order asymptotically efficient.*

In the above theorem we showed an optimal property of the exact maximum likelihood estimator. However if n is large the exact theory is intractable in practice because the likelihood function needs the inversion procedure of the $n \times n$ matrix Σ_n. Thus we often use handy 'quasi' likelihoods as approximations. Here we shall investigate an optimal property of a quasi-maximum likelihood estimator $\hat{\theta}_{qML}$ of θ, which maximizes the quasi-(log)likelihood

$$l_n(\theta) = -\frac{1}{2} \sum_{j=0}^{n-1} \left\{ \log f_\theta(\lambda_j) + \frac{I_n(\lambda_j)}{f_\theta(\lambda_j)} \right\}, \qquad (2.2.43)$$

with respect to θ, where $\lambda_j = 2\pi j/n$, and

$$I_n(\lambda_j) = \frac{1}{2\pi n} \left| \sum_{t=1}^{n} X_t e^{-it\lambda_j} \right|^2.$$

Then we shall show that an appropriately modified $\hat{\theta}_{qML}$ is second-order asymptotically efficient. We set

$$\tilde{Z}_1(\theta) = \frac{1}{\sqrt{n}} \frac{\partial}{\partial \theta} l_n(\theta),$$

$$\tilde{Z}_2(\theta) = \frac{1}{\sqrt{n}} \left[\frac{\partial^2}{\partial \theta^2} l_n(\theta) - E_\theta \left\{ \frac{\partial^2}{\partial \theta^2} l_n(\theta) \right\} \right],$$

$$B(\theta) = \frac{1}{4\pi} \int_{-\pi}^{\pi} \left\{ \frac{\partial}{\partial \theta} f_\theta(\lambda) \right\} b_\theta(\lambda) f_\theta(\lambda)^{-2} \, d\lambda,$$

$$b_\theta(\lambda) = \frac{1}{2\pi} \sum_{j=-\infty}^{\infty} |j| \gamma(j) e^{ij\lambda},$$

where $\gamma(j) = E_\theta X_t X_{t+j}$. The following lemma is useful for bias evaluation of $\hat{\theta}_{qML}$.

Lemma 2.2.2. *Suppose that Assumption 2.2.1 is satisfied, then*

$$E_\theta I_n(\lambda) = f_\theta(\lambda) - \frac{b_\theta(\lambda)}{n} + o(n^{-1}),$$

where the error term $o(n^{-1})$ is uniform in $\lambda \in [-\pi, \pi]$.

Proof. The periodogram $I_n(\lambda)$ is written in form

$$I_n(\lambda) = \frac{1}{2\pi} \sum_{j=-n+1}^{n-1} \left(1 - \frac{|j|}{n} \right) c(j) e^{ij\lambda},$$

where $c(j) = \frac{1}{n-|j|} \sum_{t=1}^{n-|j|} X_t X_{t+|j|}$. Thus

$$\begin{aligned} E_\theta I_n(\lambda) &= \frac{1}{2\pi} \sum_{j=-n+1}^{n-1} \gamma(j) e^{ij\lambda} - \frac{1}{2\pi n} \sum_{j=-n+1}^{n-1} |j| \gamma(j) e^{ij\lambda} \\ &= f_\theta(\lambda) - \frac{1}{2\pi} \sum_{|j| \geq n} \gamma(j) e^{ij\lambda} - \frac{1}{n} \left\{ b_\theta(\lambda) - \frac{1}{2\pi} \sum_{|j| \geq n} |j| \gamma(j) e^{ij\lambda} \right\} \\ &= f_\theta(\lambda) - \frac{1}{n} b_\theta(\lambda) + o(n^{-1}), \end{aligned}$$

because

$$\left| \sum_{|j| \geq n} \gamma(j) e^{ij\lambda} \right| \leq n^{-1} \sum_{|j| \geq n} |j| |\gamma(j)| = o(n^{-1}).$$

The following lemma will be used for evaluations of the asymptotic moments and stochastic expansion of $\hat{\theta}_{qML}$.

Lemma 2.2.3. *Suppose that Assumptions 2.2.1 and 2.2.2 are satisfied, then*

$$E_\theta \tilde{Z}_1(\theta) = -\frac{1}{\sqrt{n}} B(\theta) + o(n^{-\frac{1}{2}}), \tag{2.2.44}$$

$$E_\theta \{ \tilde{Z}_1(\theta)^2 \} = I(\theta) + O(n^{-1}), \tag{2.2.45}$$

$$E_\theta \{ \tilde{Z}_1(\theta) \tilde{Z}_2(\theta) \} = J(\theta) + O(n^{-1}), \tag{2.2.46}$$

$$\text{cum}_\theta \{ \tilde{Z}_1(\theta), \tilde{Z}_1(\theta), \tilde{Z}_1(\theta) \} = \frac{1}{\sqrt{n}} K(\theta) + O(n^{-\frac{3}{2}}), \tag{2.2.47}$$

$$E_\theta \left\{ \frac{1}{n} \frac{\partial^2}{\partial \theta^2} l_n(\theta) \right\} = -I(\theta) + O(n^{-1}), \tag{2.2.48}$$

$$E_\theta \left\{ \frac{1}{n} \frac{\partial^3}{\partial \theta^3} l_n(\theta) \right\} = -3J(\theta) - K(\theta) + O(n^{-1}), \tag{2.2.49}$$

and the Jth $(J \geq 3)$ order cumulant of $\tilde{Z}_{j_1}(\theta), \ldots, \tilde{Z}_{j_J}(\theta)$ $(j_1, \ldots, j_J = 1 \text{ or } 2)$ satisfies

$$\text{cum}_\theta^{(J)} \{ \tilde{Z}_{j_1}(\theta), \ldots, \tilde{Z}_{j_J}(\theta) \} = O(n^{-\frac{J}{2}+1}). \tag{2.2.50}$$

Proof. We provide only the proofs of (2.2.44), (2.2.46), (2.2.47) and (2.2.50) because the other relations can be proved similarly.

Noting Lemma 2.2.2, we have

$$
\begin{aligned}
E_\theta \tilde{Z}_1(\theta) &= -\frac{1}{2\sqrt{n}} \sum_{j=0}^{n-1} \left\{ \frac{\partial}{\partial \theta} f_\theta(\lambda_j) \cdot f_\theta(\lambda_j)^{-1} - E_\theta I_n(\lambda_j) \frac{\partial}{\partial \theta} f_\theta(\lambda_j) \cdot f_\theta(\lambda_j)^{-2} \right\} \\
&= -\frac{1}{2\sqrt{n}} \sum_{j=0}^{n-1} \left\{ \frac{\partial}{\partial \theta} f_\theta(\lambda_j) \cdot f_\theta(\lambda_j)^{-1} - (f_\theta(\lambda_j) - n^{-1} b_\theta(\lambda_j)) \frac{\partial}{\partial \theta} f_\theta(\lambda_j) \cdot f_\theta(\lambda_j)^{-2} \right\} + o(n^{-\frac{1}{2}}) \\
&= -\frac{1}{4\pi\sqrt{n}} \int_{-\pi}^{\pi} \frac{\partial}{\partial \theta} f_\theta(\lambda) b_\theta(\lambda) f_\theta(\lambda)^{-2} \, d\lambda + o(n^{-\frac{1}{2}}),
\end{aligned}
$$

which implies (2.2.44).

Since

$$
\begin{aligned}
\frac{\partial^2}{\partial \theta^2} l_n(\theta) &= -\frac{1}{2} \sum_{j=0}^{n-1} \left[\frac{\partial^2}{\partial \theta^2} f_\theta(\lambda_j) \cdot f_\theta(\lambda_j)^{-1} - \left(\frac{\partial}{\partial \theta} f_\theta(\lambda_j) \right)^2 f_\theta(\lambda_j)^{-2} \right. \\
&\quad \left. - I_n(\lambda_j) \left\{ \frac{\partial^2}{\partial \theta^2} f_\theta(\lambda_j) \cdot f_\theta(\lambda_j)^{-2} - 2 \left(\frac{\partial}{\partial \theta} f_\theta(\lambda_j) \right)^2 f_\theta(\lambda_j)^{-3} \right\} \right],
\end{aligned}
$$

we have

$$E_\theta\{\tilde{Z}_1(\theta)\tilde{Z}_2(\theta)\}$$
$$= \text{Cov}_\theta\{\tilde{Z}_1(\theta), \tilde{Z}_2(\theta)\}$$
$$= \frac{1}{4n}\sum_{j_1=0}^{n-1}\sum_{j_1=0}^{n-1}\text{Cov}_\theta\left[\frac{\partial}{\partial\theta}f_\theta(\lambda_{j_1})\cdot f_\theta(\lambda_{j_1})^{-2}\cdot I_n(\lambda_{j_1}),\right.$$
$$\left.\left\{\frac{\partial^2}{\partial\theta^2}f_\theta(\lambda_{j_2})\cdot f_\theta(\lambda_{j_2})^{-2} - 2\left(\frac{\partial}{\partial\theta}f_\theta(\lambda_{j_2})\right)^2 f_\theta(\lambda_{j_2})^{-3}\right\}I_n(\lambda_{j_2})\right]. \qquad (2.2.51)$$

Brillinger(1969) showed

$$\text{Cov}_\theta\{I_n(\lambda_{j_1}), I_n(\lambda_{j_2})\}$$
$$= n^{-2}\{|\Delta_n(\lambda_{j_1} - \lambda_{j_2})|^2 + |\Delta_n(\lambda_{j_1} + \lambda_{j_2})|^2\}f_\theta(\lambda_{j_1})^2 + n^{-2}R_n(\lambda_{j_1}, \lambda_{j_2}), \qquad (2.2.52)$$

where there is a finite $c > 0$ such that

$$|R_n(\lambda_{j_1}, \lambda_{j_2})| \leq C\{|\Delta_n(\lambda_{j_1} - \lambda_{j_2})| + |\Delta_n(\lambda_{j_1} + \lambda_{j_2})|\},$$

and $\Delta_n(\lambda) = \sum_{t=0}^{n-1}\exp(-i\lambda t)$. From (2.2.52) it is not difficult to show that (2.2.51) is

$$\frac{1}{4\pi}\sum_{j_1=0}^{n-1}\left\{\frac{\partial}{\partial\theta}f_\theta(\lambda_{j_1})\cdot f_\theta(\lambda_{j_1})^{-2}\right\}\left\{\frac{\partial^2}{\partial\theta^2}f_\theta(\lambda_{j_1})\cdot f_\theta(\lambda_{j_1})^{-2} - 2\left(\frac{\partial}{\partial\theta}f_\theta(\lambda_{j_1})\right)^2 f_\theta(\lambda_{j_1})^{-3}\right\}$$
$$\times\{2f_\theta(\lambda_{j_1})^2\} + O(n^{-1})$$
$$= \frac{1}{4\pi}\int_{-\pi}^{\pi}\left\{\frac{\partial}{\partial\theta}f_\theta(\lambda)\cdot\frac{\partial^2}{\partial\theta^2}f_\theta(\lambda)\cdot f_\theta(\lambda)^{-2} - 2\left(\frac{\partial}{\partial\theta}f_\theta(\lambda)\right)^3 f_\theta(\lambda)^{-3}\right\}d\lambda + O(n^{-1})$$

which implies (2.2.46).

Now we have

$$\text{cum}_\theta\{\tilde{Z}_1(\theta), \tilde{Z}_1(\theta), \tilde{Z}_1(\theta)\}$$
$$= \frac{1}{8n\sqrt{n}}\sum_{j_1=0}^{n-1}\sum_{j_2=0}^{n-1}\sum_{j_3=0}^{n-1}\frac{\partial}{\partial\theta}f_\theta(\lambda_{j_1})\cdot f_\theta(\lambda_{j_1})^{-2}\cdots\frac{\partial}{\partial\theta}f_\theta(\lambda_{j_3})\cdot f_\theta(\lambda_{j_3})^{-2}$$
$$\times\text{cum}_\theta\{I_n(\lambda_{j_1}), I_n(\lambda_{j_2}), I_n(\lambda_{j_3})\}. \qquad (2.2.53)$$

Define $d_n(\lambda) = \sum_{t=1}^{n}X_t e^{-i\lambda t}$. It follows from Theorem 2.3.2 in Brillinger(1975) that (2.2.53) is

$$\frac{1}{8n\sqrt{n}}\sum_{j_1=0}^{n-1}\sum_{j_2=0}^{n-1}\sum_{j_3=0}^{n-1}\frac{\partial}{\partial\theta}f_\theta(\lambda_{j_1})\cdot f_\theta(\lambda_{j_1})^{-2}\cdots\frac{\partial}{\partial\theta}f_\theta(\lambda_{j_3})\cdot f_\theta(\lambda_{j_3})^{-2}$$
$$\times\frac{1}{(2\pi n)^3}\sum_{p}^{*}\text{cum}_\theta\{d_n(\lambda_{j_{P_1}}), d_n(\lambda_{j_{P_2}})\}\cdots\text{cum}_\theta\{d_n(\lambda_{j_{P_5}}), d_n(\lambda_{j_{P_6}})\}, \qquad (2.2.54)$$

where the summation $\sum_{p}^{*}$ is over all indecomposable partitions of $\begin{pmatrix} (j_1, -j_1) \\ (j_2, -j_2) \\ (j_3, -j_3) \end{pmatrix}$. By Lemma 2.1

in Brillinger(1969a), we can see that (2.2.54) is

$$\frac{1}{n\sqrt{n}} \sum_{j_1=0}^{n-1} \{f_\theta(\lambda_{j_1})^3 + O(n^{-1})\} \left\{\frac{\partial}{\partial\theta} f_\theta(\lambda_{j_1})\right\}^3 f_\theta(\lambda_{j_1})^{-6} + O(n^{-\frac{3}{2}})$$

$$= \frac{1}{2\pi\sqrt{n}} \int_{-\pi}^{\pi} \left\{\frac{\partial}{\partial\theta} f_\theta(\lambda)\right\}^3 f_\theta(\lambda)^{-3}\, d\lambda + O(n^{-\frac{3}{2}}),$$

which implies (2.2.47).

The relation (2.2.50) follows similarly by noting that

$$\mathrm{cum}_\theta^{(J)}\{\tilde{Z}_{j_1}(\theta),\ldots,\tilde{Z}_{j_J}(\theta)\}$$

$$= \left(\frac{1}{\sqrt{n}}\right)^J \sum_{j_1=0}^{n-1} \cdots \sum_{j_J=0}^{n-1} O(1)$$

$$\times \frac{1}{(2\sqrt{n})^J} \sum_{P}^{*} \mathrm{cum}_\theta\{d_n(\lambda_{j_{P_1}}), d_n(\lambda_{j_{P_2}})\} \cdots \mathrm{cum}_\theta\{d_n(\lambda_{j_{P_{2J-1}}}), d_n(\lambda_{j_{P_{2J}}})\},$$

where the term $O(1)$ is uniform in $(j_1,\ldots,j_J)$, and $\sum_{P}^{*}$ is over all indecomposable partitions of

$$\begin{pmatrix} (j_1, -j_1) \\ \vdots \\ (j_J, -j_J) \end{pmatrix}.$$

The following theorem follows as did the proof of Theorem 2.2.3.

Theorem 2.2.5. *Under Assumptions 2.2.1 - 2.2.4,*

$$\sqrt{n}(\hat{\theta}_{qML} - \theta) = \frac{\tilde{Z}_1(\theta)}{I(\theta)} + \frac{\tilde{Z}_1(\theta)\tilde{Z}_2(\theta)}{I(\theta)^2\sqrt{n}} - \frac{3J(\theta)+K(\theta)}{2I(\theta)^3\sqrt{n}}\tilde{Z}_1(\theta)^2 + o_p(n^{-\frac{1}{2}}).$$

If we define $V_n = \sqrt{n}(\hat{\theta}_{qML} - \theta)$, from Lemma 2.2.3 and Theorem 2.2.5 we see that

$$E_\theta V_n = -\frac{B(\theta)}{I(\theta)\sqrt{n}} - \frac{J(\theta)+K(\theta)}{2I(\theta)^2\sqrt{n}} + o(n^{-\frac{1}{2}}), \tag{2.2.55}$$

$$\mathrm{cum}_\theta\{V_n, V_n\} = I(\theta)^{-1} + o(n^{-\frac{1}{2}}), \tag{2.2.56}$$

$$\mathrm{cum}_\theta\{V_n, V_n, V_n\} = -\frac{3J(\theta)+2K(\theta)}{I(\theta)^3\sqrt{n}} + o(n^{-\frac{1}{2}}), \tag{2.2.57}$$

$$\mathrm{cum}^{(J)}\{V_n\} = O(n^{-\frac{J}{2}+1}), \quad \text{for } J \geq 4. \tag{2.2.58}$$

From (2.1.16) the Edgeworth expansion for $\hat{\theta}_{qML}$ is

$$P_\theta^n\left[\sqrt{I(\theta)n}(\hat{\theta}_{qML} - \theta) \leq y\right]$$

$$= \Phi(y) - \phi(y)\left\{-\frac{1}{\sqrt{n}}\left(\frac{B(\theta)}{I(\theta)^{\frac{1}{2}}} + \frac{J(\theta)+K(\theta)}{2I(\theta)^{\frac{3}{2}}}\right) - \frac{3J(\theta)+2K(\theta)}{6I(\theta)^{\frac{3}{2}}\sqrt{n}}(y^2-1)\right\} + o(n^{-\frac{1}{2}}),$$

which implies that $\hat{\theta}_{qML}$ is not second-order AMU. If we put

$$\hat{\theta}_{qML}^* = \hat{\theta}_{qML} + \frac{B(\hat{\theta}_{qML})}{nI(\hat{\theta}_{qML})} + \frac{K(\hat{\theta}_{qML})}{6nI(\hat{\theta}_{qML})^2}$$

$$= \hat{\theta}_{qML} + \frac{B(\theta)}{nI(\theta)} + \frac{K(\theta)}{6nI(\theta)^2} + o_p(n^{-1}),$$

then we obtain

$$P_\theta^n \left[\sqrt{n}(\hat{\theta}_{qML}^* - \theta) \leq 0 \right] = \frac{1}{2} + o(n^{-\frac{1}{2}}),$$

i.e., $\hat{\theta}_{qML}^*$ is second-order AMU, and

$$P_\theta^n \left[\sqrt{n}(\hat{\theta}_{qML}^* - \theta) \leq x \right] = \Phi\left\{ x\sqrt{I(\theta)} \right\} + \frac{x^2}{6\sqrt{I(\theta)}n}\{3J(\theta) + 2K(\theta)\}\phi\left\{ x\sqrt{I(\theta)} \right\} + o(n^{-\frac{1}{2}}),$$

which coincides with the bound distribution in Theorem 2.2.2. Thus we have

Theorem 2.2.6. *The modified quasi-maximum likelihood estimator $\hat{\theta}_{qML}^*$ is second-order asymptotically efficient.*

We now proceed to calculate $I(\theta)$, $J(\theta)$, $K(\theta)$ and $B(\theta)$ for various rational spectra. This enables us to present the second-order AMU estimators for these spectra explicitly. The asymptotic bias for the maximum likelihood and quasi-maximum likelihood estimators will be evaluated.

Case 1. Consider the ARMA(p, q) spectral density

$$f_\theta(\lambda) = \frac{\sigma^2}{2\pi} \frac{\left| \sum_{j=0}^q \alpha_j e^{ij\lambda} \right|^2}{\left| \sum_{j=0}^p \beta_j e^{ij\lambda} \right|^2}.$$

Suppose that σ^2 is unknown (i.e., $\theta = \sigma^2$), and that $\alpha_0, \ldots, \alpha_q$, $\beta_0, \ldots, \beta_p$ are known. Then it is easy to show

$$I(\sigma^2) = \frac{1}{2\sigma^4}, \quad K(\sigma^2) = \frac{1}{\sigma^6}, \quad J(\sigma^2) = -\frac{1}{\sigma^6}. \tag{2.2.59}$$

Let $\hat{\sigma}_{ML}^2$ be the exact maximum likelihood estimator of σ^2. Then we can see that

$$\hat{\sigma}_{ML}^{2*} = \hat{\sigma}_{ML}^2 + \frac{K(\hat{\sigma}_{ML}^2)}{6nI(\hat{\sigma}_{ML}^2)^2} = \left(1 + \frac{2}{3n}\right)\hat{\sigma}_{ML}^2$$

is second-order AMU and efficient. Remembering (2.2.36) we have

$$E_\theta \hat{\sigma}_{ML}^2 = \sigma^2 + o(n^{-1}).$$

Case 2. Consider the ARMA(p, q) spectral density

$$f_\theta(\lambda) = \frac{\sigma^2}{2\pi} \frac{\prod_{k=1}^q (1 - \psi_k e^{i\lambda})(1 - \psi_k e^{-i\lambda})}{\prod_{k=1}^p (1 - \rho_k e^{i\lambda})(1 - \rho_k e^{-i\lambda})}, \tag{2.2.60}$$

where $\psi_1, \ldots, \psi_q$, $\rho_1, \ldots, \rho_p$ are real numbers such that $|\psi_j| < 1$, $j = 1, \ldots, q$, $|\rho_j| < 1$, $j = 1, \ldots, p$. Suppose that ψ_m is an unknown parameter (i.e., $\theta = \psi_m$), and that $\rho_1, \ldots, \rho_p$, $\psi_1, \ldots, \psi_{m-1}$, $\psi_{m+1}, \ldots, \psi_q$ are known parameters. Noting that

$$\frac{\partial}{\partial\theta} f_\theta(\lambda) \cdot f_\theta(\lambda)^{-1} = \frac{-z^2 + 2\psi_m z - 1}{(1 - \psi_m z)(z - \psi_m)},$$

where $z = e^{i\lambda}$, we have

$$I(\psi_m) = \frac{1}{4\pi i} \int_{|z|=1} \frac{(-z^2 + 2\psi_m z - 1)^2}{(1 - \psi_m z)^2 (z - \psi_m)^2 z} \, dz,$$

$$K(\psi_m) = \frac{1}{2\pi i} \int_{|z|=1} \frac{(-z^2 + 2\psi_m z - 1)^3}{(1 - \psi_m z)^3 (z - \psi_m)^3 z} \, dz.$$

These integrals are easily evaluated by the residue theorem (e.g., Hille(1959));

Theorem 2.2.7. *Suppose that $F(z)$ is holomorphic inside and on a "scroc" C, save for a finite number of isolated singularities, $a_1, \ldots, a_r$, none of which lie on C. Then*

$$\int_C F(z)\,dz = 2\pi i \sum_{j=1}^{r} \text{Res}(j),$$

where $\text{Res}(j)$ is the residue of $F(z)$ at a_j. Also if a_j is a pole of order s, then the required residue is given by

$$\text{Res}(j) = \frac{1}{(s-1)!} \left\{ \frac{d^{s-1}}{dz^{s-1}} (z-a_j)^s F(z) \right\}_{z=a_j}.$$

By this theorem we have

$$I(\psi_m) = \frac{1}{1-\psi_m^2}, \quad K(\psi_m) = \frac{-6\psi_m}{(1-\psi_m^2)^2}. \tag{2.2.61}$$

Let $\hat{\psi}_{m,ML}$ be the exact maximum likelihood estimator of ψ_m. Then the estimator

$$\hat{\psi}^*_{m,ML} = \left(1 - \frac{1}{n}\right) \hat{\psi}_{m,ML}$$

is second-order AMU and efficient. Similarly we can show that

$$J(\psi_m) = \frac{4\psi_m}{(1-\psi_m^2)^2}$$

and

$$E_\theta \hat{\psi}_{m,ML} = \psi_m + \frac{\psi_m}{n} + o(n^{-1}). \tag{2.2.62}$$

In the special case of $MA(1)$-model with the parameter ψ_1, Nishio(1981) evaluated the bias of $\hat{\psi}_{1,ML}$ up to order n^{-1}.

Case 3. We also deal with the rational spectral density (2.2.60). Assume that ρ_m is an unknown parameter (i.e., $\theta = \rho_m$), and that $\psi_1, \ldots, \psi_q, \ \rho_1, \ldots, \rho_{m-1}, \ \rho_{m+1}, \ldots, \rho_p$ are known parameters. Then

$$I(\rho_m) = \frac{1}{1-\rho_m^2}, \quad K(\rho_m) = \frac{6\rho_m}{(1-\rho_m^2)^2}, \quad J(\rho_m) = \frac{-2\rho_m}{(1-\rho_m^2)^2} \tag{2.2.63}$$

Let $\hat{\rho}_{m,ML}$ be the exact maximum likelihood estimator of ρ_m. We can see that

$$\hat{\rho}^*_{m,ML} = \left(1 + \frac{1}{n}\right) \hat{\rho}_{m,ML}$$

is second-order AMU and efficient, and that

$$E_\theta \hat{\rho}_{m,ML} = \rho_m - \frac{2\rho_m}{n} + o(n^{-1}). \tag{2.2.64}$$

Henceforth we shall consider the quasi-maximum likelihood estimation. Since the evaluation of $B(\theta)$ for general spectral density such as (2.2.60) is very complicated, we shall confine ourselves to the following ARMA$(1,1)$ spectral density;

$$f_\theta(\lambda) = \frac{\sigma^2}{2\pi} \frac{|1 - \psi e^{i\lambda}|^2}{|1 - \rho e^{i\lambda}|^2}, \tag{2.2.65}$$

where $|\psi| < 1$, $|\rho| < 1$, $\psi \neq \rho$. Then we can show that

$$\gamma(n) = \frac{\sigma^2(1 - \psi\rho)(\rho - \psi)}{(1 - \rho^2)}\rho^{n-1}, \quad \text{for } n \geq 1,$$

$$b_\theta(\lambda) = \frac{\sigma^2}{2\pi}\frac{(1 - \psi\rho)(\rho - \psi)}{(1 - \rho^2)}\frac{z\{(z^2 + 1) - 4\rho z + \rho^2(z^2 + 1)\}}{(1 - \rho z)^2(z - \rho)^2}.$$

Case 4. Suppose that σ^2 is an unknown parameter (i.e., $\theta = \sigma^2$), and that ψ and ρ are known parameters. We can show

$$B(\sigma^2) = -\frac{(\rho - \psi)^2}{\sigma^2(1 - \rho^2)(1 - \psi^2)}.$$

Let $\hat{\sigma}^2_{qML}$ be the quasi-maximum likelihood estimator of σ^2. Then

$$\hat{\sigma}^{2*}_{qML} = \hat{\sigma}^2_{qML} + \frac{2}{3n}\hat{\sigma}^2_{qML} - \frac{2}{n}\frac{(\rho - \psi)^2}{(1 - \rho^2)(1 - \psi^2)}\hat{\sigma}^2_{qML}$$

is second-order AMU and efficient. From (2.2.55) we obtain

$$E_\theta\hat{\sigma}^2_{qML} = \sigma^2 + \frac{2}{n}\frac{(\rho - \psi)^2\sigma^2}{(1 - \rho^2)(1 - \psi^2)} + o(n^{-1}).$$

Case 5. In the model (2.2.65), suppose that ρ is an unknown parameter (i.e., $\theta = \rho$), and that σ^2 and ψ are known parameters. Then it is not difficult to show

$$B(\rho) = \frac{(\rho - \psi)(1 - 2\rho\psi + \psi^2)}{(1 - \rho^2)(1 - \rho\psi)(1 - \psi^2)}.$$

Let $\hat{\rho}_{qML}$ be the quasi-maximum likelihood estimator of ρ. It follows that

$$\hat{\rho}^*_{qML} = \hat{\rho}_{qML} + \frac{1}{n}\frac{(\hat{\rho}_{qML} - \psi)(1 - 2\hat{\rho}_{qML} \cdot \psi + \psi^2)}{(1 - \hat{\rho}_{qML} \cdot \psi)(1 - \psi^2)} + \frac{1}{n}\hat{\rho}_{qML} \tag{2.2.66}$$

is second-order AMU and efficient, and that

$$E_\theta\hat{\rho}_{qML} = \rho - \frac{(\rho - \psi)(1 - 2\rho\psi + \psi^2)}{n(1 - \rho\psi)(1 - \psi^2)} - \frac{2\rho}{n} + o(n^{-1}). \tag{2.2.67}$$

Consider the case $\psi = 0$ (i.e., our model is an autoregressive model of order 1), then (2.2.66) and (2.2.67) are

$$\hat{\rho}^*_{qML} = \left(1 + \frac{2}{n}\right)\hat{\rho}_{qML}, \quad E_\theta\hat{\rho}_{qML} = \rho - \frac{3\rho}{n} + o(n^{-1}), \tag{2.2.68}$$

respectively. By the way, in the case of $\psi = 0$, we can see that $\hat{\rho}_{qML}$ is asymptotically equivalent to the Yule-Walker estimator;

$$\frac{\sum_{t=1}^{n-1} X_t X_{t+1}}{\sum_{t=1}^{n} X_t^2},$$

neglecting the terms of order $O_p(\rho^n)$, which do not disturb our asymptotic theory.

Case 6. In the model (2.2.65), suppose that ψ is an unknown parameter (i.e., $\theta = \psi$), and that σ^2 and ρ are known parameters. It is not so hard to show

$$B(\psi) = \frac{(\psi - \rho)(1 + \psi^2 - 2\psi\rho - \rho^2 + 3\psi^2\rho^2 - 2\psi^3\rho)}{(1 - \psi^2)^2(1 - \psi\rho)(1 - \rho^2)}. \tag{2.2.69}$$

Let $\hat{\psi}_{qML}$ be the quasi-maximum likelihood estimator of ψ. Then

$$\hat{\psi}_{qML}^* = \hat{\psi}_{qML} - \frac{1}{n}\hat{\psi}_{qML}$$

$$+ \frac{(\hat{\psi}_{qML} - \rho)(1 + \hat{\psi}_{qML}^2 - 2\hat{\psi}_{qML} \cdot \rho - \rho^2 + 3\hat{\psi}_{qML}^2 \cdot \rho^2 - 2\hat{\psi}_{qML}^3 \cdot \rho)}{n(1 - \hat{\psi}_{qML}^2)(1 - \hat{\psi}_{qML} \cdot \rho)(1 - \rho^2)} \tag{2.2.70}$$

is second-order AMU and efficient. Consider the case $\rho = 0$ (i.e., our model is a moving average model of order 1), then (2.2.69) and (2.2.70) are

$$B(\psi) = \frac{(1 + \psi^2)\psi}{(1 - \psi^2)^2},$$

$$\hat{\psi}_{qML}^* = \hat{\psi}_{qML} + \frac{1}{n}\frac{2\hat{\psi}_{qML}^3}{1 - \hat{\psi}_{qML}^2},$$

respectively. Also, in the case of $\rho = 0$, we have

$$E_\theta \hat{\psi}_{qML} = \psi - \frac{1}{n}\frac{2\psi^3}{1 - \psi^2} + o(n^{-1}).$$

Finally we mention that the maximum likelihood estimator is also optimal in another type of approach. For a Gaussian linear process Hosoya(1979) showed that the maximum likelihood estimator for a spectral parameter is second-order efficient in the sense of Rao(1962).

2.3. Third-order asymptotic efficiency for Gaussian ARMA processes

In this section, furthermore we develop the results of Section 2.2 for third-order case, and investigate various third-order asymptotic properties of the maximum likelihood estimators for Gaussian ARMA processes. For an AR(1) process, Fujikoshi and Ochi(1984) investigated some third-order asymptotic properties of the maximum likelihood estimator.

First we derive the third-order bound distribution for the class $\mathbf{A}_3$ of third-order AMU estimators. Then it is shown that the maximum likelihood estimators for ARMA process are not always third-order asymptotically efficient in $\mathbf{A}_3$. We also give a necessary and sufficient condition for the spectral density such that appropriately modified maximum likelihood estimator is third-order asymptotically efficient in $\mathbf{A}_3$. These results do not mean that the maximum likelihood estimator is poor in third-order sense. If we confine our discussions to an appropriate class $\mathbf{D}(\mathbf{D} \subset \mathbf{A}_3)$ of estimators, then we can show that appropriately modified maximum likelihood estimator is always third-order asymptotically efficient in $\mathbf{D}$. That is, it gives the highest probability concentration around the true value among estimators in $\mathbf{D}$.

Now we define $\mathbf{D}_2$ by

$$\mathbf{D}_2 = \left\{ f : f(\lambda) = \sum_{u=-\infty}^{\infty} a(u) \exp(-iu\lambda), \ a(u) = a(-u), \ \sum_{u=-\infty}^{\infty} |u|^2 |a(u)| < \infty \right\},$$

and set down the following assumptions.

Assumption 2.3.1. The process $\{X_t; t = 0, \pm 1, \pm 2, \dots,\}$ is a Gaussian stationary process with the spectral density $f_{\boldsymbol{\theta}}(\lambda) \in \mathbf{D}_{ARMA}$ (defined by (2.2.2)), $\boldsymbol{\theta} = (\theta_1, \dots, \theta_p)' \in \Theta \subset \mathbf{R}^p$, and mean 0.

Assumption 2.3.2. The spectral density $f_{\boldsymbol{\theta}}(\lambda)$ is continuously five times differentiable with respect to $\boldsymbol{\theta}$, and the derivatives $\partial f_{\boldsymbol{\theta}}/\partial \theta_j$, $\partial^2 f_{\boldsymbol{\theta}}/\partial \theta_j \partial \theta_k, \dots, \partial^5 f_{\boldsymbol{\theta}}/\partial \theta_j \partial \theta_k \partial \theta_m \partial \theta_l \partial \theta_r$ $(j, k, m, l, r = 1, \dots, p)$ belong to $\mathbf{D}_2$.

Assumption 2.3.3. If $\boldsymbol{\theta} \neq \boldsymbol{\theta}^*$, then $f_{\boldsymbol{\theta}} \neq f_{\boldsymbol{\theta}^*}$ on a set of positive Lebesgue measure.

Assumption 2.3.4. The matrix

$$I(\boldsymbol{\theta}) = \frac{1}{4\pi} \int_{-\pi}^{\pi} \left\{ \frac{\partial}{\partial \boldsymbol{\theta}} \log f_{\boldsymbol{\theta}}(\lambda) \right\} \left\{ \frac{\partial}{\partial \boldsymbol{\theta}'} \log f_{\boldsymbol{\theta}}(\lambda) \right\} d\lambda$$

is positive definite for all $\boldsymbol{\theta} \in \Theta$.

Suppose that a stretch, $\mathbf{X}_n = (X_1, \dots, X_n)'$ of the series $\{X_t\}$ is available. Let Σ_n be the covariance matrix of $\mathbf{X}_n$. The likelihood function based on $\mathbf{X}_n$ is given by

$$l_n(\boldsymbol{\theta}) = (2\pi)^{-\frac{n}{2}} |\Sigma_n|^{-\frac{1}{2}} \exp\left\{ -\frac{1}{2} \mathbf{X}_n' \Sigma_n^{-1} \mathbf{X}_n \right\}.$$

Let

$$Z_i = \frac{1}{\sqrt{n}} \frac{\partial}{\partial \theta_i} \log l_n(\boldsymbol{\theta}),$$

$$Z_{ij} = \frac{1}{\sqrt{n}} \left\{ \frac{\partial^2}{\partial \theta_i \partial \theta_j} \log l_n(\boldsymbol{\theta}) - E_{\boldsymbol{\theta}} \frac{\partial^2}{\partial \theta_i \partial \theta_j} \log l_n(\boldsymbol{\theta}) \right\},$$

and

$$Z_{ijk} = \frac{1}{\sqrt{n}} \left\{ \frac{\partial^3}{\partial \theta_i \partial \theta_j \partial \theta_k} \log l_n(\boldsymbol{\theta}) - E_{\boldsymbol{\theta}} \frac{\partial^3}{\partial \theta_i \partial \theta_j \partial \theta_k} \log l_n(\boldsymbol{\theta}) \right\},$$

for $i, j, k = 1, \dots, p$. Here we can see that

$$\frac{\partial}{\partial \theta_i} \log l_n(\boldsymbol{\theta}) = \frac{1}{2} \mathbf{X}_n' \Sigma_n^{-1} \Sigma_n^{(i)} \Sigma_n^{-1} \mathbf{X}_n - \frac{1}{2} \text{tr} \Sigma_n^{-1} \Sigma_n^{(i)}, \qquad (2.3.1)$$

$$\frac{\partial^2}{\partial \theta_i \partial \theta_j} \log l_n(\boldsymbol{\theta}) = -\frac{1}{2} \mathbf{X}_n' \Sigma_n^{-1} \{ \Sigma_n^{(j)} \Sigma_n^{-1} \Sigma_n^{(i)} + \Sigma_n^{(i)} \Sigma_n^{-1} \Sigma_n^{(j)} - \Sigma_n^{(i,j)} \} \Sigma_n^{-1} \mathbf{X}_n$$

$$- \frac{1}{2} \text{tr} \{ \Sigma_n^{-1} \Sigma_n^{(i,j)} - \Sigma_n^{-1} \Sigma_n^{(j)} \Sigma_n^{-1} \Sigma_n^{(i)} \}, \qquad (2.3.2)$$

and

$$\frac{\partial^3}{\partial\theta_i\partial\theta_j\partial\theta_k}\log l_n(\boldsymbol{\theta})$$

$$\begin{aligned}
= \ & \frac{1}{2}\mathbf{X}_n'\Sigma_n^{-1}\{\Sigma_n^{(k)}\Sigma_n^{-1}\Sigma_n^{(j)}\Sigma_n^{-1}\Sigma_n^{(i)} + \Sigma_n^{(j)}\Sigma_n^{-1}\Sigma_n^{(k)}\Sigma_n^{-1}\Sigma_n^{(i)} + \Sigma_n^{(j)}\Sigma_n^{-1}\Sigma_n^{(i)}\Sigma_n^{-1}\Sigma_n^{(k)} \\
& + \Sigma_n^{(k)}\Sigma_n^{-1}\Sigma_n^{(i)}\Sigma_n^{-1}\Sigma_n^{(j)} + \Sigma_n^{(i)}\Sigma_n^{-1}\Sigma_n^{(k)}\Sigma_n^{-1}\Sigma_n^{(j)} + \Sigma_n^{(i)}\Sigma_n^{-1}\Sigma_n^{(j)}\Sigma_n^{-1}\Sigma_n^{(k)} \\
& - \Sigma_n^{(j,k)}\Sigma_n^{-1}\Sigma_n^{(i)} - \Sigma_n^{(j)}\Sigma_n^{-1}\Sigma_n^{(i,k)} - \Sigma_n^{(i,k)}\Sigma_n^{-1}\Sigma_n^{(j)} - \Sigma_n^{(i)}\Sigma_n^{-1}\Sigma_n^{(j,k)} \\
& - \Sigma_n^{(k)}\Sigma_n^{-1}\Sigma_n^{(i,j)} - \Sigma_n^{(i,j)}\Sigma_n^{-1}\Sigma_n^{(k)} + \Sigma_n^{(i,j,k)}\}\Sigma_n^{-1}\mathbf{X}_n \\
& - \frac{1}{2}\mathrm{tr}\{\Sigma_n^{-1}\Sigma_n^{(k)}\Sigma_n^{-1}\Sigma_n^{(j)}\Sigma_n^{-1}\Sigma_n^{(i)} + \Sigma_n^{-1}\Sigma_n^{(j)}\Sigma_n^{-1}\Sigma_n^{(k)}\Sigma_n^{-1}\Sigma_n^{(i)} - \Sigma_n^{-1}\Sigma_n^{(k)}\Sigma_n^{-1}\Sigma_n^{(i,j)} \\
& - \Sigma_n^{-1}\Sigma_n^{(j,k)}\Sigma_n^{-1}\Sigma_n^{(i)} - \Sigma_n^{-1}\Sigma_n^{(j)}\Sigma_n^{-1}\Sigma_n^{(i,k)} + \Sigma_n^{-1}\Sigma_n^{(i,j,k)}\},
\end{aligned} \tag{2.3.3}$$

where $\Sigma_n^{(i)}$, $\Sigma_n^{(i,j)}$ and $\Sigma_n^{(i,j,k)}$ are the $n \times n$ Toeplitz type matrices whose (l, m)-th elements are given by

$$\int_{-\pi}^{\pi} e^{i(l-m)\lambda}\frac{\partial}{\partial\theta_i}f_{\boldsymbol{\theta}}(\lambda)\,d\lambda, \quad \int_{-\pi}^{\pi} e^{i(l-m)\lambda}\frac{\partial^2}{\partial\theta_i\partial\theta_j}f_{\boldsymbol{\theta}}(\lambda)\,d\lambda$$

and

$$\int_{-\pi}^{\pi} e^{i(l-m)\lambda}\frac{\partial^3}{\partial\theta_i\partial\theta_j\partial\theta_k}f_{\boldsymbol{\theta}}(\lambda)\,d\lambda, \quad \text{respectively.}$$

In the sequel we shall deal with statistics which are approximated by simple functions of Z_i, Z_{ij} and Z_{ijk}. To give their asymptotic expansions we evaluate the asymptotic cumulants (moments) of Z_i, Z_{ij} and Z_{ijk}. Using Theorem 2.2.1 and Lemma 2.2.1, we can prove

Lemma 2.3.1. *Under Assumptions 2.3.1 - 2.3.4,*

$$E(Z_iZ_j) = I_{ij} + O(n^{-1}),$$
$$E(Z_iZ_{jk}) = J_{ijk} + O(n^{-1}),$$
$$E(Z_iZ_jZ_k) = \frac{1}{\sqrt{n}}K_{ijk} + O(n^{-\frac{3}{2}}),$$
$$E(Z_iZ_{jkm}) = L_{ijkm} + O(n^{-1}),$$
$$\mathrm{Cov}(Z_{ij}, Z_{km}) = M_{ijkm} + O(n^{-1}),$$
$$E(Z_iZ_jZ_{km}) = \frac{1}{\sqrt{n}}N_{ijkm} + O(n^{-\frac{3}{2}}),$$
$$\mathrm{cum}\{Z_i, Z_j, Z_k, Z_m\} = \frac{1}{n}H_{ijkm} + O(n^{-2}),$$

where

$$I_{ij} = \frac{1}{4\pi}\int_{-\pi}^{\pi}\frac{\partial}{\partial\theta_i}\{f_{\boldsymbol{\theta}}(\lambda)\}\frac{\partial}{\partial\theta_j}\{f_{\boldsymbol{\theta}}(\lambda)\}f_{\boldsymbol{\theta}}(\lambda)^{-2}\,d\lambda,$$

$$\begin{aligned}
J_{ijk} = \ & -\frac{1}{2\pi}\int_{-\pi}^{\pi}\frac{\partial}{\partial\theta_i}\{f_{\boldsymbol{\theta}}(\lambda)\}\frac{\partial}{\partial\theta_j}\{f_{\boldsymbol{\theta}}(\lambda)\}\frac{\partial}{\partial\theta_k}\{f_{\boldsymbol{\theta}}(\lambda)\}f_{\boldsymbol{\theta}}(\lambda)^{-3}\,d\lambda \\
& + \frac{1}{4\pi}\int_{-\pi}^{\pi}\frac{\partial}{\partial\theta_i}\{f_{\boldsymbol{\theta}}(\lambda)\}\frac{\partial^2}{\partial\theta_j\partial\theta_k}\{f_{\boldsymbol{\theta}}(\lambda)\}f_{\boldsymbol{\theta}}(\lambda)^{-2}\,d\lambda,
\end{aligned}$$

$$K_{ijk} = \frac{1}{2\pi} \int_{-\pi}^{\pi} \frac{\partial}{\partial \theta_i} \{f_\theta(\lambda)\} \frac{\partial}{\partial \theta_j} \{f_\theta(\lambda)\} \frac{\partial}{\partial \theta_k} \{f_\theta(\lambda)\} f_\theta(\lambda)^{-3} \, d\lambda,$$

$$\begin{aligned}
L_{ijkm} = & \frac{3}{2\pi} \int_{-\pi}^{\pi} \frac{\partial}{\partial \theta_i} \{f_\theta(\lambda)\} \frac{\partial}{\partial \theta_j} \{f_\theta(\lambda)\} \frac{\partial}{\partial \theta_k} \{f_\theta(\lambda)\} \frac{\partial}{\partial \theta_m} \{f_\theta(\lambda)\} f_\theta(\lambda)^{-4} \, d\lambda \\
& - \frac{1}{2\pi} \int_{-\pi}^{\pi} \frac{\partial}{\partial \theta_i} \{f_\theta(\lambda)\} \left[\frac{\partial^2}{\partial \theta_k \partial \theta_m} \{f_\theta(\lambda)\} \frac{\partial}{\partial \theta_j} \{f_\theta(\lambda)\} \right. \\
& \left. + \frac{\partial^2}{\partial \theta_j \partial \theta_k} \{f_\theta(\lambda)\} \frac{\partial}{\partial \theta_m} \{f_\theta(\lambda)\} + \frac{\partial^2}{\partial \theta_j \partial \theta_m} \{f_\theta(\lambda)\} \frac{\partial}{\partial \theta_k} \{f_\theta(\lambda)\} \right] f_\theta(\lambda)^{-3} \, d\lambda \\
& + \frac{1}{4\pi} \int_{-\pi}^{\pi} \frac{\partial}{\partial \theta_i} \{f_\theta(\lambda)\} \frac{\partial^3}{\partial \theta_j \partial \theta_k \partial \theta_k} \{f_\theta(\lambda)\} f_\theta(\lambda)^{-2} \, d\lambda,
\end{aligned}$$

$$\begin{aligned}
M_{ijkm} = & \frac{1}{\pi} \int_{-\pi}^{\pi} \frac{\partial}{\partial \theta_i} \{f_\theta(\lambda)\} \frac{\partial}{\partial \theta_j} \{f_\theta(\lambda)\} \frac{\partial}{\partial \theta_k} \{f_\theta(\lambda)\} \frac{\partial}{\partial \theta_m} \{f_\theta(\lambda)\} f_\theta(\lambda)^{-4} \, d\lambda \\
& - \frac{1}{2\pi} \int_{-\pi}^{\pi} \left[\frac{\partial}{\partial \theta_i} \{f_\theta(\lambda)\} \frac{\partial}{\partial \theta_j} \{f_\theta(\lambda)\} \frac{\partial^2}{\partial \theta_k \partial \theta_m} \{f_\theta(\lambda)\} \right. \\
& \left. + \frac{\partial}{\partial \theta_k} \{f_\theta(\lambda)\} \frac{\partial}{\partial \theta_m} \{f_\theta(\lambda)\} \frac{\partial^2}{\partial \theta_i \partial \theta_j} \{f_\theta(\lambda)\} \right] f_\theta(\lambda)^{-3} \, d\lambda \\
& + \frac{1}{4\pi} \int_{-\pi}^{\pi} \frac{\partial^2}{\partial \theta_i \partial \theta_j} \{f_\theta(\lambda)\} \frac{\partial^2}{\partial \theta_k \partial \theta_m} \{f_\theta(\lambda)\} f_\theta(\lambda)^{-2} \, d\lambda,
\end{aligned}$$

$$\begin{aligned}
N_{ijkm} = & -\frac{1}{\pi} \int_{-\pi}^{\pi} \frac{\partial}{\partial \theta_i} \{f_\theta(\lambda)\} \frac{\partial}{\partial \theta_j} \{f_\theta(\lambda)\} \frac{\partial}{\partial \theta_k} \{f_\theta(\lambda)\} \frac{\partial}{\partial \theta_m} \{f_\theta(\lambda)\} f_\theta(\lambda)^{-4} \, d\lambda \\
& + \frac{1}{2\pi} \int_{-\pi}^{\pi} \frac{\partial}{\partial \theta_i} \{f_\theta(\lambda)\} \frac{\partial}{\partial \theta_j} \{f_\theta(\lambda)\} \frac{\partial^2}{\partial \theta_k \partial \theta_m} \{f_\theta(\lambda)\} f_\theta(\lambda)^{-3} \, d\lambda,
\end{aligned}$$

$$H_{ijkm} = \frac{3}{2\pi} \int_{-\pi}^{\pi} \frac{\partial}{\partial \theta_i} \{f_\theta(\lambda)\} \frac{\partial}{\partial \theta_j} \{f_\theta(\lambda)\} \frac{\partial}{\partial \theta_k} \{f_\theta(\lambda)\} \frac{\partial}{\partial \theta_m} \{f_\theta(\lambda)\} f_\theta(\lambda)^{-4} \, d\lambda.$$

Henceforth if θ is scalar we use I, J, K, etc. (or $I(\theta), J(\theta), K(\theta)$, etc.) instead of I_{ij}, J_{ijk}, K_{ijk}, etc. for simplicity.

We turn next to the derivation of the third-order bound distribution for $\mathbf{A}_3$. Denote the log-likelihood function based on $\mathbf{X}_n$ by

$$G_n(\theta) = -\frac{n}{2} \log 2\pi - \frac{1}{2} \log |\Sigma_n| - \frac{1}{2} \mathbf{X}_n' \Sigma_n^{-1} \mathbf{X}. \tag{2.3.4}$$

Consider the problem of testing hypothesis $H : \theta = \theta_0 + x/\sqrt{n}$ $(x > 0)$ against alternative $K : \theta = \theta_0$. Let $LR = G_n(\theta_0) - G_n(\theta_1)$, where $\theta_1 = \theta_0 + x/\sqrt{n}$. To give the bound distribution (see (2.1.5)), we derive the Edgeworth expansion of LR. Since the spectral density $f_\theta(\lambda)$ is continuously five times differentiable we get

$$\begin{aligned}
LR = & -\frac{x}{\sqrt{n}} \left\{ \frac{\partial}{\partial \theta} G_n(\theta) \right\}_{\theta_0} - \frac{x^2}{2n} \left\{ \frac{\partial^2}{\partial \theta^2} G_n(\theta) \right\}_{\theta_0} - \frac{x^3}{6n\sqrt{n}} \left\{ \frac{\partial^3}{\partial \theta^3} G_n(\theta) \right\}_{\theta_0} \\
& - \frac{x^4}{24n^2} \left\{ \frac{\partial^4}{\partial \theta^4} G_n(\theta) \right\}_{\theta_0} - \frac{x^5}{120n^2\sqrt{n}} \left\{ \frac{\partial^5}{\partial \theta^5} G_n(\theta) \right\}_{\theta'},
\end{aligned}$$

where $\theta_0 \leq \theta' \leq \theta_1$. The derivatives of $G_n(\theta)$ can be written as $\frac{\partial^j}{\partial \theta^j} G_n(\theta) = \mathbf{X}'_n A_j \mathbf{X}_n + \mathrm{tr} B_j$, $j = 1, \ldots, 5$, where A_j and B_j are of the form $\Lambda_1^{-1} \Gamma_2 \Lambda_2^{-1} \cdots \Gamma_s \Lambda_s^{-1}$ defined in Theorem 2.2.1. We write

$$E_\theta \left\{ \frac{1}{\sqrt{n}} \frac{\partial}{\partial \theta} G_n(\theta) \right\}^2 = I(\theta) + \frac{\Delta(\theta)}{n} + o(n^{-1}),$$

where $\Delta(\theta)$ will be explicitly evaluated in the case of ARMA(1,1). In order to derive the Edgeworth expansions of LR under $\theta = \theta_0$ and $\theta = \theta_1$, we have only to evaluate the asymptotic cumulants (moments) of LR under $\theta = \theta_0$ and $\theta = \theta_1$. Using Theorem 2.2.1, Lemma 2.2.1 these cumulants (moments) are given by the manner used in (2.2.15) – (2.2.21);

$$E_{\theta_0}(LR) = \frac{x^2}{2} I(\theta_0) + \frac{x^3}{6\sqrt{n}} \{3J(\theta_0) + K(\theta_0)\} + \frac{x^2}{2n} \Delta(\theta_0)$$
$$+ \frac{x^4}{24n} \{4L(\theta_0) + 3M(\theta_0) + 6N(\theta_0) + H(\theta_0)\} + o(n^{-1}), \tag{2.3.5}$$

$$E_{\theta_0}(LR - E_{\theta_0} LR)^2 = x^2 I(\theta_0) + \frac{x^3}{\sqrt{n}} J(\theta_0) + \frac{x^4}{4n} M(\theta_0) + \frac{x^4}{3n} L(\theta_0)$$
$$+ \frac{x^2}{n} \Delta(\theta_0) + o(n^{-1}), \tag{2.3.6}$$

$$\mathrm{cum}_{\theta_0} \{LR, LR, LR\} = -\frac{x^3}{\sqrt{n}} K(\theta_0) - \frac{3x^4}{2n} N(\theta_0) + o(n^{-1}), \tag{2.3.7}$$

$$\mathrm{cum}_{\theta_0} \{LR, LR, LR, LR\} = \frac{x^4}{n} H(\theta_0) + o(n^{-1}), \tag{2.3.8}$$

$$\mathrm{cum}_{\theta_0}^{(J)} \{LR, \ldots, LR\} = O(n^{-\frac{J}{2}+1}), \quad \text{for } J \geq 5, \tag{2.3.9}$$

where $\mathrm{cum}_{\theta_0}^{(J)} \{LR, \ldots, LR\}$ is the Jth-order cumulant of LR. Putting $V_n = \{LR - E_{\theta_1}(LR)\}/\{x\sqrt{I_n}\}$, where $I_n = I + \Delta/n$, similarly we have

$$\mathrm{Var}_{\theta_1}(V_n) = 1 + \frac{1}{\sqrt{n}} b_1 + \frac{1}{n} b_2 + o(n^{-1}), \tag{2.3.10}$$

$$\mathrm{cum}_{\theta_1} \{V_n, V_n, V_n\} = \frac{1}{\sqrt{n}} c_1 + \frac{1}{n} c_2 + o(n^{-1}), \tag{2.3.11}$$

$$\mathrm{cum}_{\theta_1} \{V_n, V_n, V_n, V_n\} = \frac{1}{n} d_1 + o(n^{-1}), \tag{2.3.12}$$

$$\mathrm{cum}_{\theta_1}^{(J)} \{V_n, \ldots, V_n\} = O(n^{-\frac{J}{2}+1}), \quad J \geq 5, \tag{2.3.13}$$

where $b_1 = x(J + K)/I$, $b_2 = x^2(4L + 3M + 18N + 6H)/12I$, $c_1 = -K/I^{3/2}$, $c_2 = -x(3N + 2H)/2I^{3/2}$, $d_1 = H/I^2$. Remembering (2.1.16) we get the Edgeworth expansion;

$$P_{\theta_1}^n[V_n \leq a] = \Phi(a) - \phi(a) \left[\frac{1}{2} \left(\frac{b_1}{\sqrt{n}} + \frac{b_2}{n} \right) a + \left(\frac{c_1}{6\sqrt{n}} + \frac{c_2}{6n} \right) (a^2 - 1) \right.$$
$$+ \left(\frac{d_1}{24n} + \frac{b_1^2}{8n} \right) (a^3 - 3a) + \frac{b_1 c_1}{12n} (a^4 - 6a^2 + 3)$$
$$\left. + \frac{c_1^2}{72n} (a^5 - 10a^3 + 15a) \right] + o(n^{-1}). \tag{2.3.14}$$

Noting $\Phi(a) = (1/2) + a\phi(a) - (a^3/2)\phi(a) + \ldots$, and if we put $a = -c_1/(6\sqrt{n}) - c_2/(6n) + b_1 c_1/(6n)$ in (2.3.14), it is easy to show

$$P_{\theta_1}^n[V_n \le a] = \frac{1}{2} + o(n^{-1}),$$ (2.3.15)

$$P_{\theta_1}^n[V_n \ge a] = \frac{1}{2} + o(n^{-1}).$$ (2.3.16)

Putting $W_n = -\{V_n - a - x\sqrt{I_n}\}$ and $x' = x\sqrt{I_n}$, similarly we get the Edgeworth expansion;

$$
\begin{aligned}
P_{\theta_0}^n[V_n \ge a] &= P_{\theta_0}^n[W_n \le x'] \\
&= \Phi(x') - \phi(x') \left\{ \frac{\beta_3}{6\sqrt{n}} + \frac{\beta_3}{6\sqrt{n}}(x'^2 - 1) \right. \\
&\quad + \left(\frac{\beta_2}{2n} + \frac{\beta_3^2}{72n} \right) x' + \left(\frac{\beta_4}{24n} + \frac{\beta_3^2}{36n} \right)(x'^3 - 3x') \\
&\quad \left. + \frac{\beta_3^2}{72n}(x'^5 - 10x'^3 + 15x') \right\} + o(n^{-1}),
\end{aligned}
$$ (2.3.17)

where $\beta_3 = -(3J + 2K)/I^{3/2}$, $\beta_4 = 3\beta_3^2 - (4L + 3M + 12N + 3H)/I^2$,

$$\beta_2 = \frac{17}{36}\beta_3^2 - \frac{K^2}{18I^3} - \frac{12L + 9M + 36N + 8H}{12I^2}.$$

If $\{\hat{\theta}_n\}$ is third-order AMU, then remembering (2.1.7), the fundamental lemma of Neyman and Pearson and $\Phi(x') = \Phi(x\sqrt{I}) + \frac{x\Delta}{2n\sqrt{I}}\phi(x\sqrt{I}) + o(n^{-1})$, we have

Theorem 2.3.1. *For any $\hat{\theta}_n \in \mathbf{A}_3$, we have*

$$\limsup_{n \to \infty} n \left[P_{\theta_0}^n \left\{ \sqrt{nI}(\hat{\theta}_n - \theta_0) \le y \right\} - F_{\theta_0}^{(3)}(y) \right] \le 0, \quad \text{for} \quad y > 0,$$ (2.3.18)

where

$$
\begin{aligned}
F_{\theta_0}^{(3)}(y) &= \Phi(y) - \phi(y) \left\{ \frac{\beta_3}{6\sqrt{n}} + \frac{\beta_3}{6\sqrt{n}}(y^2 - 1) + \left(\frac{\beta_2}{2n} + \frac{\beta_3^2}{72n} - \frac{\Delta}{2In} \right) y \right. \\
&\quad \left. + \left(\frac{\beta_4}{24n} + \frac{\beta_3^2}{36n} \right)(y^3 - 3y) + \frac{\beta_3^2}{72n}(y^5 - 10y^3 + 15y) \right\}.
\end{aligned}
$$

For $y < 0$, similarly we have

$$\liminf_{n \to \infty} n \left[P_{\theta_0}^n \left\{ \sqrt{nI}(\hat{\theta}_n - \theta_0) \le y \right\} - F_{\theta_0}^{(3)}(y) \right] \ge 0.$$ (2.3.19)

Now we seek the bound distribution $F_0^{(3)}(y)$ for concrete parametrization of the spectral density. Calculations for I, J, K were already given by the residue theorem (see (2.2.59), (2.2.61), (2.2.63)). However direct calculations for $\Delta(\theta)$ are very troublesome (i.e., n^{-1}-order term of $n^{-1}\mathrm{tr}\Sigma_n^{-1}\dot{\Sigma}_n\Sigma_n^{-1}\dot{\Sigma}_n$). Thus we make the following device. The first part of the following lemma is given by differentiating $\log \det \Sigma_n$ twice. The second part is essentially due to Galbraith and Galbraith(1974).

Lemma 2.3.2. *Suppose that the spectral density $f_\theta(\lambda)$ of $\{X_t\}$ is given by*

$$f_\theta(\lambda) = \frac{\sigma^2}{2\pi} \frac{|1 - \beta e^{i\lambda}|^2}{|1 - \alpha e^{i\lambda}|^2}.$$ (2.3.20)

Then we have

$$\mathrm{tr}\Sigma_n^{-1}\dot{\Sigma}_n\Sigma_n^{-1}\dot{\Sigma}_n = -\frac{\partial^2}{\partial\theta^2}\log\det\Sigma_n + \mathrm{tr}\Sigma_n^{-1}\ddot{\Sigma}_n,$$

$$\log\det\Sigma_n = 2\log(1-\alpha\beta) - \log(1-\alpha^2) - \log(1-\beta^2) + n\log\sigma^2 + O(\beta^{2n}),$$

where $\dot{\Sigma}_n$ and $\ddot{\Sigma}_n$ are the $n \times n$ Toeplitz type matrices whose (m,l)-th elements are given by $\int_{-\pi}^{\pi} e^{i(m-l)\lambda}\frac{\partial}{\partial\theta}f_\theta(\lambda)\,d\lambda$ and $\int_{-\pi}^{\pi} e^{i(m-l)\lambda}\frac{\partial^2}{\partial\theta^2}f_\theta(\lambda)\,d\lambda$, respectively.

Put $\Sigma_n^{-1} = \{m_{rs}\}$, $r,s = 1,\ldots,n$. Galbraith and Galbraith(1974) gave the exact expressions of m_{rs} for the ARMA(1,1) process with the spectral density (2.3.20). From their exact expressions we get

$$
\begin{aligned}
m_{rr} &= \frac{(\beta-\alpha)^2}{\sigma^2(1-\beta^2)}[1 - \beta^{2(r-1)} - \beta^{2(n-r)}] + \frac{1}{\sigma^2} + O(\beta^n), \\
&= \tilde{m}_{rr} + O(\beta^n), \quad (\text{say}), \quad r = 1,\ldots,n, \\
m_{rs} &= \frac{\beta^{s-r-1}(\beta-\alpha)(1-\alpha\beta)}{\sigma^2(1-\beta^2)} - \frac{(\beta-\alpha)^2}{\sigma^2(1-\beta^2)}\beta^{r+s-2} - \frac{(\beta-\alpha)^2}{\sigma^2(1-\beta^2)}\beta^{2n-s-r} + O(\beta^n), \\
&= \tilde{m}_{rs} + O(\beta^n), \quad (\text{say}), \quad 1 \le r < s \le n.
\end{aligned}
$$

Then

$$\mathrm{tr}\Sigma_n^{-1}\ddot{\Sigma}_n = \sum_{r=1}^{n}\tilde{m}_{rr}a_{rr} + 2\sum_{r=1}^{n-1}\sum_{s=r+1}^{n}\tilde{m}_{rs}a_{rs} + O(n^2\beta^n), \tag{2.3.21}$$

where a_{rs} is the (r,s)-th element of $\ddot{\Sigma}_n$. Using the residue theorem, Lemma 2.3.2 and (2.3.21) we have

Proposition 2.3.1. For the spectral density

$$f_\theta(\lambda) = \frac{\sigma^2}{2\pi}\frac{|1-\beta e^{i\lambda}|^2}{|1-\alpha e^{i\lambda}|^2},$$

we have

$$
\begin{aligned}
L(\sigma^2) &= \frac{3}{\sigma^8}, & L(\beta) &= \frac{6+18\beta^2}{(1-\beta^2)^3}, & L(\alpha) &= 0, \\
M(\sigma^2) &= \frac{2}{\sigma^8}, & M(\beta) &= \frac{6+14\beta^2}{(1-\beta^2)^3}, & M(\alpha) &= \frac{2+2\alpha^2}{(1-\alpha^2)^3}, \\
N(\sigma^2) &= -\frac{2}{\sigma^8}, & N(\beta) &= \frac{-8-20\beta^2}{(1-\beta^2)^3}, & N(\alpha) &= \frac{-4-8\alpha^2}{(1-\alpha^2)^3}, \\
H(\sigma^2) &= \frac{3}{\sigma^8}, & H(\beta) &= \frac{6(3+7\beta^2)}{(1-\beta^2)^3}, & H(\alpha) &= \frac{6(3+7\alpha^2)}{(1-\alpha^2)^3}, \\
\Delta(\sigma^2) &= 0, & \Delta(\beta) &= \frac{\alpha^2(1-5\beta^2)+\alpha(4\beta+4\beta^3)-3\beta^2-1}{(1-\beta^2)^2(1-\alpha\beta)^2}, \\
\Delta(\alpha) &= \frac{\beta^2(3\alpha^2-1)-4\alpha^3\beta+3\alpha^2-1}{(1-\alpha^2)^2(1-\alpha\beta)^2}. &&
\end{aligned}
$$

Using the above results we have

Theorem 2.3.2. *For the ARMA spectral density model*

$$f_\theta(\lambda) = \frac{\sigma^2}{2\pi} \frac{|1 - \beta e^{i\lambda}|^2}{|1 - \alpha e^{i\lambda}|^2},$$

the third-order bound distributions for $\theta = \sigma^2, \beta$ and α are given by

$$F_{\sigma^2}^{(3)}(y) = \Phi(y) - \phi(y)\left[\frac{\sqrt{2}y^2}{3\sqrt{n}} + \frac{5y}{18n} - \frac{7y^3}{18n} + \frac{y^5}{9n}\right],$$

$$F_{\beta}^{(3)}(y) = \Phi(y) - \phi(y)\left[\frac{\alpha^2(3\beta^4 + 13\beta^2 - 2) - 2\alpha(7\beta^3 + 7\beta) + 9\beta^2 + 5}{4n(1-\beta^2)(1-\alpha\beta)^2}\right]y,$$

$$F_{\alpha}^{(3)}(y) = \Phi(y) - \phi(y)\left[-\frac{\alpha y^2}{\sqrt{n(1-\alpha^2)}} + \frac{\{(3\alpha^4 - 3\alpha^2 + 2)\beta^2 + (2\alpha^3 - 6\alpha)\beta + 5 - 3\alpha^2\}y}{4n(1-\alpha\beta)^2(1-\alpha^2)}\right.$$
$$\left. - \frac{(1+2\alpha^2)y^3}{2n(1-\alpha^2)} + \frac{\alpha^2 y^5}{2n(1-\alpha^2)}\right].$$

Remark 2.3.1. In the special case of $\beta = 0$, i.e., an autoregressive model of order 1, the above bound distribution $F_{\alpha}^{(3)}(y)$ becomes

$$\Phi(y) + \phi(y)\left[\frac{\alpha y^2}{\sqrt{n(1-\alpha^2)}} + \frac{(3\alpha^2 - 5)y}{4n(1-\alpha^2)} + \frac{(1+2\alpha^2)y^3}{2n(1-\alpha^2)} - \frac{\alpha^2 y^5}{2n(1-\alpha^2)}\right],$$

which coincides with the result of Fujikoshi and Ochi(1984).

Next we investigate the third-order asymptotic properties of the maximum likelihood estimators. It will be shown that appropriately modified (to be third-order AMU) maximum likelihood estimators are not always third-order asymptotically efficient in $\mathbf{A}_3$.

We first set down the notations;

$$Z^{(1)}(\theta) = \frac{1}{\sqrt{n}}\frac{\partial}{\partial\theta}G_n(\theta),$$

$$Z^{(2)}(\theta) = \frac{1}{\sqrt{n}}\left\{\frac{\partial^2}{\partial\theta^2}G_n(\theta) - E_\theta\frac{\partial^2}{\partial\theta^2}G_n(\theta)\right\},$$

$$Z^{(3)}(\theta) = \frac{1}{\sqrt{n}}\left\{\frac{\partial^3}{\partial\theta^3}G_n(\theta) - E_\theta\frac{\partial^3}{\partial\theta^3}G_n(\theta)\right\},$$

where $G_n(\theta)$ is defined by (2.3.4). For simplicity we sometimes use $Z^{(1)}, Z^{(2)}, Z^{(3)}$ instead of $Z^{(1)}(\theta), Z^{(2)}(\theta), Z^{(3)}(\theta)$, respectively. Notice that

$$E_\theta n^{-1}\frac{\partial^3}{\partial\theta^3}G_n(\theta) = -3J(\theta) - K(\theta) + O(n^{-1}), \tag{2.3.22}$$

$$E_\theta n^{-1}\frac{\partial^4}{\partial\theta^4}G_n(\theta) = -4L(\theta) - 3M(\theta) - 6N(\theta) - H(\theta) + O(n^{-1}). \tag{2.3.23}$$

Using (2.3.22) and (2.3.23) we proceed as in the proof of Theorem 2.2.3 to derive the stochastic expansion of the maximum likelihood estimator $\hat{\theta}_{ML}$ of θ:

Theorem 2.3.3. *Under Assumptions 2.3.1 - 2.3.4,*

$$
\sqrt{n}(\hat{\theta}_{ML} - \theta) = \frac{Z^{(1)}}{I_n} + \frac{1}{\sqrt{n}I^2}\left\{ Z^{(1)}Z^{(2)} - \frac{3J + K}{2I}Z^{(1)^2} \right\}
$$
$$
+ \frac{1}{nI^3}\left\{ Z^{(1)}Z^{(2)^2} + \frac{1}{2}Z^{(1)^2}Z^{(3)} - \frac{3(3J + K)}{2I}Z^{(1)^2}Z^{(2)} \right.
$$
$$
\left. + \frac{(3J + K)^2}{2I^2}Z^{(1)^3} - \frac{4L + 3M + 6N + H}{6I}Z^{(1)^3} \right\} + o_p(n^{-1}).
$$

Let $U_n = \sqrt{n}(\hat{\theta}_{ML} - \theta)$. By Lemma 2.3.1 and Theorem 2.3.3 we can show that

$$
E_\theta U_n = -\frac{J + K}{2\sqrt{n}I^2} + o(n^{-1}), \tag{2.3.24}
$$
$$
\mathrm{Var}_\theta(U_n) = I^{-1} - \frac{\Delta}{I^2 n} + \frac{7J^2 + 14JK + 5K^2}{2I^4 n} - \frac{L + 4N + H}{I^3 n} + o(n^{-1}), \tag{2.3.25}
$$
$$
\mathrm{cum}_\theta\{U_n, U_n, U_n\} = -\frac{3J + 2K}{I^3\sqrt{n}} + o(n^{-1}), \tag{2.3.26}
$$
$$
\mathrm{cum}_\theta\{U_n, U_n, U_n, U_n\} = \frac{12(2J + K)(J + K)}{I^5 n} - \frac{4L + 12N + 3H}{I^4 n} + o(n^{-1}), \tag{2.3.27}
$$
$$
\mathrm{cum}_\theta^{(J)}\{U_n, \ldots, U_n\} = O(n^{-\frac{J}{2}+1}), \quad \text{for } J \geq 5, \tag{2.3.28}
$$

which imply

Theorem 2.3.4. *Under Assumptions 2.3.1 - 2.3.4,*

$$
P_\theta^n\left[\sqrt{nI}(\hat{\theta}_{ML} - \theta) \leq x \right]
$$
$$
= \Phi(x) - \phi(x)\left\{ \frac{\alpha_1}{\sqrt{n}} + \frac{\gamma_1}{6\sqrt{n}}(x^2 - 1) + \frac{1}{2}(\frac{\rho_2}{n} + \frac{\alpha_1^2}{n})x \right.
$$
$$
\left. + (\frac{\delta_1}{24n} + \frac{\alpha_1\gamma_1}{6n})(x^3 - 3x) + \frac{\gamma_1^2}{72n}(x^5 - 10x^3 + 15x) \right\} + o(n^{-1}), \tag{2.3.29}
$$

where

$$
\alpha_1 = -\frac{J + K}{2I^{\frac{3}{2}}}, \quad \rho_2 = \frac{7J^2 + 14JK + 5K^2}{2I^3} - \frac{L + 4N + H}{I^2} - \frac{\Delta}{I},
$$

$$
\gamma_1 = -\frac{3J + 2K}{I^{\frac{3}{2}}}, \quad \delta_1 = \frac{12(2J + K)(J + K)}{I^3} - \frac{4L + 12N + 3H}{I^2}.
$$

In the special case of

$$
f_\theta(\lambda) = \frac{\sigma^2}{2\pi}\frac{|1 - \beta e^{i\lambda}|^2}{|1 - \alpha e^{i\lambda}|^2},
$$

we have

$$
P_\theta^n\left[\sqrt{\frac{n}{2\sigma^4}}(\hat{\sigma}_{ML}^2 - \sigma^2) \leq x \right]
$$
$$
= \Phi(x) - \phi(x)\left\{ \frac{\sqrt{2}}{3\sqrt{n}}(x^2 - 1) + \frac{x}{6n} - \frac{11x^3}{18n} + \frac{x^5}{9n} \right\} + o(n^{-1}), \tag{2.3.30}
$$

$$P_\theta^n \left[\sqrt{\frac{n}{1-\beta^2}} (\hat{\beta}_{ML} - \beta) \leq x \right]$$

$$= \Phi(x) - \phi(x) \left\{ \frac{\beta}{\sqrt{n}\sqrt{1-\beta^2}} + \frac{\alpha^2(\beta^4 + 17\beta^2 - 2) - \alpha(22\beta + 10\beta^3) + 7\beta^2 + 9}{4n(1-\beta^2)(1-\alpha\beta)^2} x \right.$$

$$\left. + \frac{3 - \beta^2}{4n(1-\beta^2)} x^3 \right\} + o(n^{-1}), \tag{2.3.31}$$

$$P_\theta^n \left[\sqrt{\frac{n}{1-\alpha^2}} (\hat{\alpha}_{ML} - \alpha) \leq x \right]$$

$$= \Phi(x) - \phi(x) \left\{ \frac{-\alpha(x^2 + 1)}{\sqrt{n}\sqrt{1-\alpha^2}} + \frac{\beta^2(\alpha^4 - 7\alpha^2 + 2) + \beta(6\alpha^3 + 2\alpha) + 1 - 5\alpha^2}{4n(1-\alpha^2)(1-\alpha\beta)^2} x \right.$$

$$\left. - \frac{\alpha^2 + 1}{4n(1-\alpha^2)} x^3 + \frac{\alpha^2 x^5}{2n(1-\alpha^2)} \right\} + o(n^{-1}), \tag{2.3.32}$$

where $\hat{\sigma}_{ML}^2$, $\hat{\beta}_{ML}$ and $\hat{\alpha}_{ML}$ are the maximum likelihood estimators of σ^2, β and α, respectively.

Remark 2.3.2. In the special case of $\beta = 0$, i.e., an autoregressive model of order 1, the right-hand side of (2.3.32) becomes

$$\Phi(x) - \phi(x) \left\{ -\frac{\alpha(x^2 + 1)}{\sqrt{n}\sqrt{1-\alpha^2}} + \frac{1 - 5\alpha^2}{4n(1-\alpha^2)} x - \frac{\alpha^2 + 1}{4n(1-\alpha^2)} x^3 + \frac{\alpha^2 x^5}{2n(1-\alpha^2)} \right\} + o(n^{-1}), \tag{2.3.33}$$

which coincides with the result of Fujikoshi and Ochi(1984).

We can also evaluate the mean square error of $\hat{\theta}_{ML}$ up to n^{-1}-order.

Theorem 2.3.5. *Under Assumptions 2.3.1 - 2.3.4,*

$$E_\theta \left\{ \sqrt{nI}(\hat{\theta}_{ML} - \theta) \right\}^2$$

$$= 1 + \frac{1}{n} \left\{ \frac{15J^2 + 30JK + 11K^2}{4I^3} - \frac{L + 4N + H}{I^2} - \frac{\Delta}{I} \right\} + o(n^{-1}). \tag{2.3.34}$$

In the special case of

$$f_\theta(\lambda) = \frac{\sigma^2}{2\pi} \frac{|1 - \beta e^{i\lambda}|^2}{|1 - \alpha e^{i\lambda}|^2},$$

we have

$$E_\theta \left\{ \sqrt{\frac{n}{2\sigma^4}} (\hat{\sigma}_{ML}^2 - \sigma^2) \right\}^2 = 1 + o(n^{-1}), \tag{2.3.35}$$

$$E_\theta \left\{ \sqrt{\frac{n}{1-\beta^2}} (\hat{\beta}_{ML} - \beta) \right\}^2$$

$$= 1 + \frac{1}{n} \left\{ \frac{\alpha^2(-\beta^4 + 13\beta^2 - 1) + \alpha(-2\beta^3 - 20\beta) + 2\beta^2 + 9}{(1-\beta^2)(1-\alpha\beta)^2} \right\} + o(n^{-1}), \tag{2.3.36}$$

$$E_\theta \left\{ \sqrt{\frac{n}{1-\alpha^2}} (\hat{\alpha}_{ML} - \alpha) \right\}^2$$

$$= 1 + \frac{1}{n} \left\{ \frac{\beta^2(14\alpha^4 - 5\alpha^2 + 1) + \beta(4\alpha - 24\alpha^3) + 11\alpha^2 - 1}{(1-\alpha^2)(1-\alpha\beta)^2} \right\} + o(n^{-1}). \qquad (2.3.37)$$

For i.i.d. multinomial case the evaluation of the type (2.3.34) for the variance of the maximum likelihood estimator has been studied earlier by Rao(1962) who introduced the concept of the second-order efficiency which corresponds to our third-order efficiency in a class **D** which will be defined later in this section.

Now we discuss the third-order asymptotic efficiency of the maximum likelihood estimators in **A**$_3$. For that purpose we modify $\hat{\theta}_{ML}$ to be third-order AMU. That is, we put

$$\hat{\theta}^*_{ML} = \hat{\theta}_{ML} + \frac{K(\hat{\theta}_{ML})}{6nI^2(\hat{\theta}_{ML})}.$$

Since

$$\sqrt{n}(\hat{\theta}^*_{ML} - \theta) = \sqrt{n}(\hat{\theta}_{ML} - \theta) + \frac{K}{6\sqrt{n}I^2} + \frac{\sqrt{n}(\hat{\theta}_{ML} - \theta)}{6n} \frac{\partial}{\partial\theta} \left\{ \frac{K}{I^2} \right\} + o_p(n^{-1}),$$

we can show that

$$n \, \mathrm{Var}(\hat{\theta}^*_{ML}) = \left\{ 1 + \frac{(3N+H)I - 2K(2J+K)}{3nI^3} \right\} n \, \mathrm{Var}(\hat{\theta}_{ML}) + o(n^{-1}). \qquad (2.3.38)$$

Putting $U^*_n = \sqrt{nI}(\hat{\theta}^*_{ML} - \theta)$, we have

$$E_\theta\{U^*_n\} = -\frac{3J+2K}{6I^{\frac{3}{2}}\sqrt{n}} + o(n^{-1}), \qquad (2.3.39)$$

$$\mathrm{Var}_\theta\{U^*_n\} = 1 - \frac{\Delta}{In} + \frac{-3L - 9N - 2H}{3I^2 n} + \frac{21J^2 + 34JK + 11K^2}{6I^3 n} + o(n^{-1}), \qquad (2.3.40)$$

$$\mathrm{cum}_\theta\{U^*_n, U^*_n, U^*_n\} = -\frac{3J+2K}{I^{\frac{3}{2}}\sqrt{n}} + o(n^{-1}), \qquad (2.3.41)$$

$$\mathrm{cum}_\theta\{U^*_n, U^*_n, U^*_n, U^*_n\} = \frac{12(2J+K)(J+K)}{I^3 n} - \frac{4L + 12N + 3H}{I^2 n} + o(n^{-1}) \qquad (2.3.42)$$

$$\mathrm{cum}_\theta^{(J)}\{U^*_n, \ldots, U^*_n\} = O(n^{-\frac{J}{2}+1}), \quad \text{for} \quad J \geq 5, \qquad (2.3.43)$$

which yield

Theorem 2.3.6. *Under Assumptions 2.3.1 - 2.3.4,*

$$P_\theta^n \left[\sqrt{nI}(\hat{\theta}^*_{ML} - \theta) \leq y \right]$$

$$= \Phi(y) - \phi(y) \left[-\frac{3J+2K}{6\sqrt{n}I^{\frac{3}{2}}} y^2 + \frac{1}{n} \left\{ -\frac{\Delta}{2I} - \frac{K^2}{36I^3} + \frac{H}{24I^2} \right\} y \right.$$

$$\left. + \frac{1}{n} \left\{ \frac{3KJ + K^2}{18I^3} - \frac{4L + 12N + 3H}{24I^2} \right\} y^3 + \frac{(3J+2K)^2}{72nI^3} y^5 \right] + o(n^{-1}). \qquad (2.3.44)$$

Thus the difference between the third-order bound distribution $F_\theta^{(3)}(y)$ given by (2.3.18) and (2.3.44) is

$$F_\theta^{(3)}(y) - P_\theta^n\left[\sqrt{nI}(\hat{\theta}_{ML}^* - \theta) \le y\right] = \frac{y^3\phi(y)}{8nI^3}\{MI - J^2\} + o(n^{-1}), \quad for \ y > 0. \qquad (2.3.45)$$

In the special case of

$$f_\theta(\lambda) = \frac{\sigma^2}{2\pi}\frac{|1 - \beta e^{i\lambda}|^2}{|1 - \alpha e^{i\lambda}|^2},$$

for $\hat{\sigma}_{ML}^{2*} = (1 + \frac{2}{3n})\hat{\sigma}_{ML}^2$, $\hat{\beta}_{ML}^* = (1 - \frac{1}{n})\hat{\beta}_{ML}$ and $\hat{\alpha}_{ML}^* = (1 + \frac{1}{n})\hat{\alpha}_{ML}$ which are third-order AMU, we have

$$F_{\sigma^2}^{(3)}(y) - P_{\sigma^2}^n\left[\sqrt{\frac{n}{2\sigma^4}}(\hat{\sigma}_{ML}^{2*} - \sigma^2) \le y\right] = o(n^{-1}), \qquad (2.3.46)$$

(i.e., $\hat{\sigma}_{ML}^{2*}$ is third-order asymptotically efficient),

$$F_\beta^{(3)}(y) - P_\beta^n\left[\sqrt{\frac{n}{1 - \beta^2}}(\hat{\beta}_{ML}^* - \beta) \le y\right] = \frac{3 - \beta^3}{4(1 - \beta^2)n}\phi(y)y^3 + o(n^{-1}), \quad for \ y > 0, \qquad (2.3.47)$$

(i.e., $\hat{\beta}_{ML}^*$ is not third-order asymptotically efficient),

$$F_\alpha^{(3)}(y) - P_\alpha^n\left[\sqrt{\frac{n}{1 - \alpha^2}}(\hat{\alpha}_{ML}^* - \alpha) \le y\right] = \frac{y^3\phi(y)}{4n} + o(n^{-1}), \quad for \ y > 0, \qquad (2.3.48)$$

(i.e., $\hat{\alpha}_{ML}^*$ is not third-order asymptotically efficient).

We also have the following:

Theorem 2.3.7. *Under Assumptions 2.3.1 – 2.3.4, the modified maximum likelihood estimator*

$$\hat{\theta}_{ML}^* = \hat{\theta}_{ML} + \frac{K(\hat{\theta}_{ML})}{6nI(\hat{\theta}_{ML})^2}$$

is third-order asymptotically efficient in $\mathbf{A}_3$ if and only if the spectral density $f_\theta(\lambda)$ satisfies the differential equation;

$$\frac{\partial^2 \log f_\theta(\lambda)}{\partial\theta^2} - \left(\frac{\partial \log f_\theta(\lambda)}{\partial\theta}\right)^2 + C(\theta)\left(\frac{\partial \log f_\theta(\lambda)}{\partial\theta}\right) = 0, \qquad (2.3.49)$$

where $C(\theta)$ is a function which depends only on θ. The condition (2.3.49) is equivalent that the spectral density $f_\theta(\lambda)$ is parametrized as

$$f_\theta(\lambda) = s(\lambda)\exp\left[\int \frac{\exp\left(\int -C(\theta)\,d\theta\right)}{-\int \exp\left(\int -C(\theta)\,d\theta\right)\,d\theta + b(\lambda)}\,d\theta\right], \qquad (2.3.50)$$

where $s(\cdot)$ and $b(\cdot)$ are functions which depend only on λ.

Proof. By (2.3.45) we can see that $\hat{\theta}_{ML}^*$ is third-order asymptotically efficient if and only if $MI - J^2 = 0$.

By Schwarz's inequality we have

$$MI = \frac{1}{4\pi} \int_{-\pi}^{\pi} \left\{ 2 \left(\frac{\partial f_\theta}{\partial \theta} \right)^2 f_\theta^{-2} - \frac{\partial^2 f_\theta}{\partial \theta^2} f_\theta^{-1} \right\}^2 d\lambda \times \frac{1}{4\pi} \int_{-\pi}^{\pi} \left(\frac{\partial f_\theta}{\partial \theta} \right)^2 f_\theta^{-2} d\lambda$$

$$\geq \left[\frac{1}{4\pi} \int_{-\pi}^{\pi} \left\{ \frac{\partial^2 f_\theta}{\partial \theta^2} f_\theta^{-1} - 2 \left(\frac{\partial f_\theta}{\partial \theta} \right)^2 f_\theta^{-2} \right\} \left\{ \frac{\partial f_\theta}{\partial \theta} f_\theta^{-1} \right\} d\lambda \right]^2 = J^2,$$

where the equality holds if and only if

$$\frac{\partial^2 f_\theta}{\partial \theta^2} f_\theta^{-1} - 2 \left(\frac{\partial f_\theta}{\partial \theta} \right)^2 f_\theta^{-2} = -C(\theta) \frac{\partial f_\theta}{\partial \theta} f_\theta^{-1}, \tag{2.3.51}$$

where $C(\theta)$ depends only on θ. The above (2.3.51) implies (2.3.49). From (2.3.49) we have

$$\frac{\partial \log f_\theta(\lambda)}{\partial \theta} = \left\{ \exp \int -C(\theta) \, d\theta \right\} \left\{ \int -e^{\int -C(\theta) \, d\theta} \, d\theta + b(\lambda) \right\}^{-1},$$

which implies (2.3.50).

As we saw in Theorem 2.3.6 the maximum likelihood estimator is not always third-order asymptotically efficient in the class $\mathbf{A}_3$.
However this result does not mean that the maximum likelihood estimator is poor in third-order sense. If we confine our discussions to a class of estimators $\mathbf{D}(\subset \mathbf{A}_3)$, then we can show that the maximum likelihood estimator which is modified to be third-order AMU, is third-order asymptotically efficient in $\mathbf{D}$ in the sense that it gives the highest probability concentration around the true value among estimators in $\mathbf{D}$. Here we shall develop our discussions for the case when the unknown parameter θ is vector-valued.

Let us remember that the process $\{X_t\}$ satisfies Assumptions 2.3.1–2.3.4. We set

$$\mathbf{Z}^{(1)} = (Z_1, \ldots, Z_p)', \quad \mathbf{Z}^{(2)} = (Z_{ij}), \quad \mathbf{U} = (U_1, \ldots, U_p)' = I(\boldsymbol{\theta})^{-1} \mathbf{Z}^{(1)}$$

and

$$E_{\boldsymbol{\theta}} \mathbf{Z}^{(1)} \mathbf{Z}^{(1)\prime} = I(\boldsymbol{\theta}) + \frac{\Delta(\boldsymbol{\theta})}{n} + o(n^{-1}).$$

Let I^{ij} and n_{ij} be the (i, j)-th elements of $I(\boldsymbol{\theta})^{-1}$ and $I(\boldsymbol{\theta})^{-1} \Delta(\boldsymbol{\theta}) I(\boldsymbol{\theta})^{-1}$, respectively, and define the differential operators:

$$D_j = \sum_{k=1}^{p} I^{jk} \frac{\partial}{\partial \theta_k}, \quad j = 1, \ldots, p.$$

Suppose that $A(\boldsymbol{\theta})$ is a measurable function of $\mathbf{X}_n = (X_1, \ldots, X_n)'$ and $\boldsymbol{\theta}$, and is differentiable with respect to $\boldsymbol{\theta}$. Calculation of $D_j E_{\boldsymbol{\theta}} A(\boldsymbol{\theta})$ yields the following lemma which is essentially due to Takeuchi(1981).

Lemma 2.3.3.

$$E_{\boldsymbol{\theta}} A(\boldsymbol{\theta}) U_j = \frac{1}{\sqrt{n}} D_j E_{\boldsymbol{\theta}} A(\boldsymbol{\theta}) - \frac{1}{\sqrt{n}} E_{\boldsymbol{\theta}} D_j A(\boldsymbol{\theta}).$$

Let **S** be the class of the estimators $\hat{\boldsymbol{\theta}}_n = (\hat{\theta}_1, \ldots, \hat{\theta}_p)'$ which are asymptotically expanded as

$$\sqrt{n}(\hat{\boldsymbol{\theta}}_n - \boldsymbol{\theta}) = \mathbf{U} + \frac{\mathbf{Q}}{\sqrt{n}} + o_p(n^{-\frac{1}{2}}), \qquad (2.3.52)$$

where $\mathbf{Q} = (Q_1, \ldots, Q_p)' = O_p(1)$. We assume that $\sqrt{n}(\hat{\boldsymbol{\theta}}_n - \boldsymbol{\theta})$ has the Edgeworth expansion up to the order n^{-1} and that

$$E_{\boldsymbol{\theta}}\sqrt{n}(\hat{\boldsymbol{\theta}}_n - \boldsymbol{\theta}) = \frac{\mu}{\sqrt{n}} + o(n^{-1}), \qquad (2.3.53)$$

where $\mu = (\mu_1, \ldots, \mu_p)' = E_{\boldsymbol{\theta}}\mathbf{Q}$. Putting

$$\mathbf{S} = (S_1, \ldots, S_p)' = \sqrt{n}(\hat{\boldsymbol{\theta}}_n - E_{\boldsymbol{\theta}}\hat{\boldsymbol{\theta}}_n),$$

we have

$$
\begin{aligned}
E_{\boldsymbol{\theta}}S_i S_j &= E_{\boldsymbol{\theta}}\{U_i + (S_i - U_i)\}\{U_j + (S_j - U_j)\} \\
&= -E_{\boldsymbol{\theta}}U_i U_j + E_{\boldsymbol{\theta}}S_i U_j + E_{\boldsymbol{\theta}}S_j U_i + E_{\boldsymbol{\theta}}(S_i - U_i)(S_j - U_j) \\
&= -E_{\boldsymbol{\theta}}U_i U_j + E_{\boldsymbol{\theta}}S_i U_j + E_{\boldsymbol{\theta}}S_j U_i + \frac{1}{n}\mathrm{Cov}_{\boldsymbol{\theta}}(Q_i, Q_j) + o(n^{-1}).
\end{aligned}
\qquad (2.3.54)
$$

Notice that

$$
\begin{aligned}
D_j S_i &= -D_j\left\{\sqrt{n}E_{\boldsymbol{\theta}}(\hat{\theta}_i)\right\} \\
&= -D_j\left\{\sqrt{n}\theta_i + \frac{\mu_i}{\sqrt{n}}\right\} + o(n^{-\frac{1}{2}}) \\
&= -\sqrt{n}I^{ji} - \frac{1}{\sqrt{n}}D_j\mu_i + o(n^{-\frac{1}{2}}).
\end{aligned}
\qquad (2.3.55)
$$

From Lemma 2.3.3 it follows that

$$E_{\boldsymbol{\theta}}S_i D_j = I^{ji} + \frac{1}{n}D_j\mu_i + o(n^{-1}). \qquad (2.3.56)$$

Then we can find

$$E_{\boldsymbol{\theta}}S_i S_j = I^{ij} - \frac{1}{n}\eta_{ij} + \frac{1}{n}D_j\mu_i + \frac{1}{n}D_i\mu_j + \frac{1}{n}\mathrm{Cov}_{\boldsymbol{\theta}}(Q_i, Q_j) + o(n^{-1}). \qquad (2.3.57)$$

Set

$$\beta_{ijk} = -\sum_{i'=1}^{p}\sum_{j'=1}^{p}\sum_{k'=1}^{p} I^{ii'}I^{jj'}I^{kk'}\{2K_{i'j'k'} + J_{i'j'k'} + J_{j'k'i'} + J_{k'i'j'}\},$$

$$A_{ijk} = E_{\boldsymbol{\theta}}U_i\tilde{Q}_j\tilde{Q}_k + E_{\boldsymbol{\theta}}U_j\tilde{Q}_i\tilde{Q}_k + E_{\boldsymbol{\theta}}U_k\tilde{Q}_i\tilde{Q}_j,$$

where $\tilde{Q}_i = Q_i - \mu_i$. Similarly we can show that

$$E_{\boldsymbol{\theta}}S_i S_j S_k = \frac{1}{\sqrt{n}}\beta_{ijk} + \frac{1}{2n}A_{ijk} + o(n^{-1}). \qquad (2.3.58)$$

We define

$$
\beta_{ijkm} = -3H^{ijkm} - 2(N^{ijkm} + N^{ikjm} + N^{imjk} + N^{jkim} + N^{jmik} + N^{kmij})
$$

$$
-(L^{ijkm} + L^{jikm} + L^{kijm} + L^{mijk}) + \frac{1}{2}(\Gamma^{ijkm} + \Gamma^{jikm} + \Gamma^{ikjm}
$$

$$
+\Gamma^{kijm} + \Gamma^{jmik} + \Gamma^{mjik} + \Gamma^{kmij} + \Gamma^{mkij} + \Gamma^{jkim} + \Gamma^{kjim} + \Gamma^{imjk} + \Gamma^{mijk}),
$$

where

$$
\Gamma^{ijkm} = \sum_{i'=1}^{p}\sum_{j'=1}^{p}\sum_{k'=1}^{p}\sum_{m'=1}^{p}\sum_{l'=1}^{p}\sum_{l=1}^{p} I^{ii'} I^{jj'} I^{kk'} I^{mm'} I^{ll'}
$$

$$
\times (K_{j'l'i'} + J_{l'i'j'} + J_{j'i'l'})(2K_{lk'm'} + J_{k'lm'} + J_{m'lk'}),
$$

$$
H^{ijkm} = \sum_{i'=1}^{p}\sum_{j'=1}^{p}\sum_{k'=1}^{p}\sum_{m'=1}^{p} I^{ii'} I^{jj'} I^{kk'} I^{mm'} H_{i'j'k'm'},
$$

$$
N^{ijkm} = \sum_{i'=1}^{p}\sum_{j'=1}^{p}\sum_{k'=1}^{p}\sum_{m'=1}^{p} I^{ii'} I^{jj'} I^{kk'} I^{mm'} N_{i'j'k'm'},
$$

$$
L^{ijkm} = \sum_{i'=1}^{p}\sum_{j'=1}^{p}\sum_{k'=1}^{p}\sum_{m'=1}^{p} I^{ii'} I^{jj'} I^{kk'} I^{mm'} L_{i'j'k'm'}.
$$

Then we can also show

$$
\text{cum}_{\boldsymbol{\theta}}(S_i, S_j, S_k, S_m) = \frac{1}{n}\beta_{ijkm} + o(n^{-1}). \tag{2.3.59}
$$

Noting (2.3.53), (2.3.57), (2.3.58) and (2.3.59) we have

Theorem 2.3.8. *For* $\hat{\boldsymbol{\theta}}_n \in \mathbf{S}$

$$
P_{\boldsymbol{\theta}}^n \left[\sqrt{n}(\hat{\boldsymbol{\theta}}_n - \boldsymbol{\theta}) \in C \right]
$$

$$
= \int \cdots \int_C N(\mathbf{y}; I(\boldsymbol{\theta})^{-1}) \left[1 + \sum_i \frac{\mu_i}{\sqrt{n}} H_i(\mathbf{y}) \right.
$$

$$
+ \frac{1}{2n}\sum_i\sum_j \{D_i\mu_j + D_j\mu_i - \eta_{ij} + \text{Cov}(Q_i, Q_j) + \mu_i\mu_j\} H_{ij}(\mathbf{y})
$$

$$
+ \sum_i\sum_j\sum_k \left\{ \frac{\beta_{ijk}}{6\sqrt{n}} + \frac{A_{ijk}}{12n} \right\} H_{ijk}(\mathbf{y}) + \sum_i\sum_j\sum_k\sum_m \left\{ \frac{\beta_{ijkm}}{24n} + \frac{\mu_i\beta_{jkm}}{6n} \right\} H_{ijkm}(\mathbf{y})
$$

$$
\left. + \frac{1}{72n}\sum_i\sum_j\sum_k\sum_{i'}\sum_{j'}\sum_{k'} \beta_{ijk}\beta_{i'j'k'} H_{ijki'j'k'}(\mathbf{y}) \right] d\mathbf{y} + o(n^{-1}), \tag{2.3.60}
$$

where C *is a convex set in* R^p.

Now we introduce a class $\mathbf{D}$ $(\subset \mathbf{S})$ of estimators which satisfy

$$
A_{ijk} = o(1) \quad \text{for} \quad i, j, k = 1, \ldots, p. \tag{2.3.61}
$$

This class $\mathbf{D}$ is a natural one. It will be shown that the maximum likelihood estimator and a quasi-maximum likelihood estimator belong to $\mathbf{D}$. First we can easily extend the result of Theorem 2.2.3 to the vector case:

Proposition 2.3.2. The maximum likelihood estimator $\hat{\boldsymbol{\theta}}_{ML}$ of $\boldsymbol{\theta}$ has the stochastic expansion

$$\sqrt{n}(\hat{\boldsymbol{\theta}}_{ML} - \boldsymbol{\theta}) = \mathbf{U} + \frac{1}{\sqrt{n}}I(\boldsymbol{\theta})^{-1}\mathbf{Z}^{(2)}\mathbf{U} + \frac{1}{2\sqrt{n}}I(\boldsymbol{\theta})^{-1}R\ldots\circ\mathbf{U}\circ\mathbf{U} + o_p(n^{-\frac{1}{2}})$$

$$= \mathbf{W}_n^{(2)} + o_p(n^{-\frac{1}{2}}),$$

say, where $R\ldots = \{R_{ijk}, \ i,j,k = 1,\ldots,p\}$, $R_{ijk} = -K_{ijk} - J_{ijk} - J_{jki} - J_{kij}$, and $R\ldots\circ\mathbf{U}\circ\mathbf{U}$ is a p-dimensional column vector with ith component $\sum_j \sum_k R_{ijk}U_jU_k$.

Thus we can express the term Q_i for $\hat{\boldsymbol{\theta}}_{ML}$ as a linear combination of $Z_{i_1}Z_{i_2}$ and $Z_{i_1}Z_{i_2i_3}$. Using the fundamental properties of the cumulants we can see that A_{ijk} is a linear combination of $\text{cum}\{Z_{i_1}, Z_{i_2}Z_{i_3}, Z_{i_4}Z_{i_5}\}$, $\text{cum}\{Z_{i_1}, Z_{i_2}Z_{i_3}, Z_{i_4}Z_{i_5i_6}\}$ and $\text{cum}\{Z_{i_1}, Z_{i_2}Z_{i_3i_4}, Z_{i_5}Z_{i_6i_7}\}$.
In view of Theorem 2.3.2 in Brillinger(1975), these typical terms are at most of order $O(n^{-\frac{1}{2}})$. We may now state

Proposition 2.3.3. The maximum likelihood estimator $\hat{\boldsymbol{\theta}}_{ML}$ of $\boldsymbol{\theta}$ belongs to $\mathbf{D}$.

We next define a quasi-maximum likelihood estimator $\hat{\boldsymbol{\theta}}_{qML}$ as an estimator which maximizes

$$-\frac{1}{2}\sum_{j=0}^{n-1}\left\{\log f_{\boldsymbol{\theta}}(\lambda_j) + \frac{IP_n(\lambda_j)}{f_{\boldsymbol{\theta}}(\lambda_j)}\right\}, \tag{2.3.62}$$

with respect to $\boldsymbol{\theta}$, where $\lambda_j = 2\pi j/n$ and

$$IP_n(\lambda_j) = \frac{1}{2\pi n}\left|\sum_{t=1}^n X_t e^{-i\lambda_j t}\right|^2.$$

Then we have

Proposition 2.3.4. The quasi-maximum likelihood estimator $\hat{\boldsymbol{\theta}}_{qML}$ belongs to $\mathbf{D}$.

The proof of this proposition is omitted because we will give the third-order Edgeworth expansion of the quasi-maximum likelihood estimator in Chapter 3, which confirms this assertion.

In view of (2.3.60), if we modify $\hat{\boldsymbol{\theta}}_n \in \mathbf{D}$ to be coordinate-wise third-order AMU, then $\boldsymbol{\mu} = (\mu_1,\ldots,\mu_p)'$ is specified by β_{ijk}. Thus, for $\hat{\boldsymbol{\theta}}_n \in \mathbf{D}\cap\mathbf{A}_3$, the terms depending on $\hat{\boldsymbol{\theta}}_n$ in (2.3.60) are only $\text{Cov}(Q_i, Q_j)$. If we get an estimator $\hat{\boldsymbol{\theta}}_n^* \in \mathbf{D}\cap\mathbf{A}_3$ which minimizes the matrix $\{\text{Cov}(Q_i, Q_j)\}$, then we can show that it maximizes the concentration probability in the sense that

$$\lim_{n\to\infty} n\left[P_{\boldsymbol{\theta}}^n\left\{\sqrt{n}I(\boldsymbol{\theta})^{\frac{1}{2}}(\hat{\boldsymbol{\theta}}_n^* - \boldsymbol{\theta}) \in C\right\} - P_{\boldsymbol{\theta}}^n\left\{\sqrt{n}I(\boldsymbol{\theta})^{\frac{1}{2}}(\hat{\boldsymbol{\theta}}_n - \boldsymbol{\theta}) \in C\right\}\right] \geq 0, \tag{2.3.63}$$

for any other $\hat{\boldsymbol{\theta}}_n \in \mathbf{D}\cap\mathbf{A}_3$ and any symmetric (about the origin) convex set C in R^p. Then we say that $\hat{\boldsymbol{\theta}}_n^*$ is third-order asymptotically efficient in $\mathbf{D}$.

Proposition 2.3.5. Suppose that an estimator $\hat{\boldsymbol{\theta}}_n$ belongs to $\mathbf{D}$. The matrix $\{\text{Cov}(Q_i, Q_j), \ i,j = 1,\ldots,p\}$ is minimized if $\sqrt{n}(\hat{\boldsymbol{\theta}}_n - \boldsymbol{\theta})$ has the stochastic expansion:

$$\sqrt{n}(\hat{\boldsymbol{\theta}}_n - \boldsymbol{\theta}) = \mathbf{W}_n^{(2)} + \frac{1}{\sqrt{n}}\boldsymbol{\xi} + o_p(n^{-\frac{1}{2}}), \tag{2.3.64}$$

where $\mathbf{W}_n^{(2)}$ is defined in Proposition 2.3.2 and $\boldsymbol{\xi}$ is a constant vector.

Proof. To make the idea clear we prove the assertion for the case where θ is scalar (i.e., $p = 1$). Here we use the scalar notations $Z^{(1)}$, $Z^{(2)}$, Q and S instead of $\mathbf{Z}^{(1)}$, $\mathbf{Z}^{(2)}$, $\mathbf{Q}$ and $\mathbf{S}$, respectively, and set $\tilde{Q} = Q - EQ$. Then

$$
\begin{aligned}
E_\theta(Z^{(1)^2}\tilde{Q}) &= \sqrt{n}E_\theta\left\{Z^{(1)^2}(S - \frac{Z^{(1)}}{I})\right\} + o(1) \\
&= \sqrt{n}E_\theta(Z^{(1)^2}S) - \frac{K}{I} + o(1).
\end{aligned}
$$

In view of Lemma 2.3.3, we have

$$
E_\theta(Z^{(1)^2}S) = \frac{\partial}{\partial\theta}\frac{E_\theta(Z^{(1)}S)}{\sqrt{n}} - E_\theta\frac{\frac{\partial}{\partial\theta}S \cdot Z^{(1)} + S \cdot \frac{\partial}{\partial\theta}Z^{(1)}}{\sqrt{n}}.
$$

Use of (2.3.55), (2.3.56) and $\frac{\partial}{\partial\theta}Z^{(1)} = Z^{(2)} - \sqrt{n}I + O(n^{-\frac{1}{2}})$ yields

$$
\begin{aligned}
E_\theta(Z^{(1)^2}S) &= -E_\theta\frac{Z^{(1)}\frac{\partial}{\partial\theta}Z^{(1)}}{\sqrt{n}I} + o(n^{-\frac{1}{2}}) \\
&= -E_\theta\frac{Z^{(1)}Z^{(2)}}{\sqrt{n}I} + o(n^{-\frac{1}{2}}) \\
&= -\frac{J}{\sqrt{n}I} + o(n^{-\frac{1}{2}}).
\end{aligned}
$$

Therefore

$$
E_\theta(Z^{(1)^2}\tilde{Q}) = -\frac{J + K}{I} + o(1). \tag{2.3.65}
$$

Similarly we can show

$$
E_\theta(Z^{(1)}Z^{(2)}\tilde{Q}) = \frac{M}{I} - \frac{J(2J + K)}{I^2} + o(1). \tag{2.3.66}
$$

In order to minimize $E_\theta\tilde{Q}^2$ under the conditions (2.3.65) and (2.3.66), $\tilde{Q}$ must be expressed as

$$
\tilde{Q} = \lambda_1(Z^{(1)}Z^{(2)} - E_\theta Z^{(1)}Z^{(2)}) + \lambda_2(Z^{(1)^2} - E_\theta Z^{(1)^2}), \tag{2.3.67}
$$

where λ_1 and λ_2 are constants. Then

$$
E_\theta(Z^{(1)^2}\tilde{Q}) = 2\lambda_1 IJ + 2\lambda_2 I^2 + o(1), \tag{2.3.68}
$$
$$
E_\theta(Z^{(1)}Z^{(2)}\tilde{Q}) = \lambda_1(IM + J^2) + 2\lambda_2 IJ + o(1). \tag{2.3.69}
$$

Combining (2.3.65) and (2.3.66) we have

$$
\lambda_1 = \frac{1}{I^2}, \quad \lambda_2 = -\frac{3J + K}{2I^3},
$$

which gives the stochastic expansion (2.3.64) for the scalar case.

From Proposition 2.3.2 the maximum likelihood estimator has the stochastic expansion (2.3.64) with $\boldsymbol{\xi} = \mathbf{O}$. Thus we have the following theorem which describes a third-order optimal property of the maximum likelihood estimator.

Theorem 2.3.9. *If we modify the maximum likelihood estimator of θ to be third-order AMU, then it is third-order asymptotically efficient in* **D**.

2.4. Normalizing transformations of some statistics of Gaussian ARMA processes

In the area of multivariate analysis several authors have considered transformations of statistics which are based upon functions of the elements of sample covariance matrix, and derived the Edgeworth expansions of the transformed statistics. Konishi(1978) gave a transformation of the sample correlation coefficient which extinguishes a part of the second-order terms of the Edgeworth expansion. Also, Konishi(1981) discussed the transformations of a statistic based upon the elements of the sample covariance matrix which extinguish the second-order terms of the Edgeworth expansions. Furthermore Fang and Krishnaiah(1982a) gave the Edgeworth expansions of certain functions of the elements of noncentral Wishart matrix; they also obtained analogous results for functions of the elements of the sample covariance matrix when the underlying distribution is a mixture of multivariate distributions.

In the area of time series analysis the first study of higher order asymptotic properties of a transformed statistic is Phillips(1979). He gave the Edgeworth expansion of a transformation of the least squares estimator for the coefficient of an AR(1) process, and showed that Fisher's z-transformation extinguishes a part of the second-order terms of the Edgeworth expansion.

Here we investigate Edgeworth type expansions of certain transformations of some statistics for Gaussian ARMA processes. We also seek transformations which will make the second-order part of the Edgeworth expansions vanish. Some numerical studies are made and they show that the above transformations give better approximations than the usual approximation. The main results of this section are based on Taniguchi, Krishnaiah and Chao(1989).

Suppose that $\{X_t; t = 0, \pm 1, \pm 2, \ldots\}$ is a Gaussian stationary process with the spectral density $f_\theta(\lambda)$. Throughout this section we assume that our process satisfies Assumptions 2.2.1 – 2.2.4, and use the notations in Section 2.2. Let $g(\theta)$ be a three times continuously differentiable function, and let $\hat{\theta}_{ML}$ be the maximum likelihood estimator of θ based on $\mathbf{X}_n = (X_1, \ldots, X_n)'$. We write

$$V_n^{(1)} = \sqrt{nI(\theta)}\frac{g(\hat{\theta}_{ML}) - g(\theta) - c/n}{g'(\theta)},$$

where c is a constant. First we give the Edgeworth expansion of V_n. Using Taylor's formula we have

$$V_n^{(1)} = \sqrt{nI(\theta)}\frac{(\hat{\theta}_{ML} - \theta)g'(\theta) + \frac{1}{2}(\hat{\theta}_{ML} - \theta)^2 g''(\theta) - c/n}{g'(\theta)} + o_p(n^{-\frac{1}{2}}). \tag{2.4.1}$$

We have already evaluated the asymptotic cumulants of $\sqrt{n}(\hat{\theta}_{ML} - \theta)$ in (2.2.36) – (2.2.39). Thus it follows that

$$EV_n^{(1)} = -\frac{J(\theta) + K(\theta)}{2\sqrt{n}I(\theta)^{\frac{3}{2}}} + \frac{g''(\theta)}{2\sqrt{n}I(\theta)^{\frac{1}{2}}g'(\theta)} - \frac{cI(\theta)^{\frac{1}{2}}}{\sqrt{n}g'(\theta)} + o(n^{-\frac{1}{2}}), \tag{2.4.2}$$

$$\text{cum}_\theta\{V_n^{(1)}, V_n^{(1)}\} = 1 + o(n^{-\frac{1}{2}}), \tag{2.4.3}$$

$$\text{cum}_\theta\{V_n^{(1)}, V_n^{(1)}, V_n^{(1)}\} = -\frac{3J(\theta) + 2K(\theta)}{\sqrt{n}I(\theta)^{\frac{3}{2}}} + \frac{3g''(\theta)}{\sqrt{n}g'(\theta)I(\theta)^{\frac{1}{2}}} + o(n^{-\frac{1}{2}}), \tag{2.4.4}$$

$$\text{cum}_\theta^{(J)}\{V_n^{(1)}\} = O(n^{-\frac{J}{2}+1}), \quad \text{for } J \geq 4. \tag{2.4.5}$$

In view of (2.1.16) we have

Theorem 2.4.1. *Under Assumptions 2.2.1 – 2.2.4,*

$$P_\theta^n \left[\sqrt{nI(\theta)} \frac{g(\hat{\theta}_{ML}) - g(\theta) - c/n}{g'(\theta)} \le x \right]$$

$$= \Phi(x) - \phi(x) \left[\frac{1}{6\sqrt{n}} \left\{ -\frac{3J(\theta) + 2K(\theta)}{I(\theta)^{\frac{3}{2}}} + \frac{3g''(\theta)}{g'(\theta)I(\theta)^{\frac{1}{2}}} \right\} x^2 \right.$$

$$\left. + \frac{1}{\sqrt{n}} \left\{ -\frac{K(\theta)}{6I(\theta)^{\frac{3}{2}}} - \frac{cI(\theta)^{\frac{1}{2}}}{g'(\theta)} \right\} \right] + o(n^{-\frac{1}{2}}). \tag{2.4.6}$$

Setting the coefficients of the second-order term in (2.4.6) equal to zero, we have

Corollary 2.4.1. Under Assumptions 2.2.1 –2.2.4, if $g_0(\theta)$ and c_0 satisfy

$$\frac{g_0''(\theta)}{g_0'(\theta)} = \frac{3J(\theta) + 2K(\theta)}{3I(\theta)} \tag{2.4.7}$$

and

$$c_0 = -\frac{K(\theta)g'(\theta)}{6I(\theta)^2} \tag{2.4.8}$$

then

$$P_\theta^n \left[\sqrt{nI(\theta)} \frac{g_0(\hat{\theta}_{ML}) - g_0(\theta) - c_0/n}{g_0'(\theta)} \le x \right] = \Phi(x) + o(n^{-\frac{1}{2}}). \tag{2.4.9}$$

In the sense of (2.4.9) the function $g_0(\theta)$ is called a normalizing transformation of $\hat{\theta}_{ML}$.

Now we seek the normalizing transformation $g_0(\cdot)$ and the constant c_0 for the ARMA spectral density

$$f_\theta(\lambda) = \frac{\sigma^2}{2\pi} \frac{\prod_{j=1}^q (1 - \psi_j e^{i\lambda})(1 - \psi_j e^{-i\lambda})}{\prod_{j=1}^p (1 - \rho_j e^{i\lambda})(1 - \rho_j e^{-i\lambda})}, \tag{2.4.10}$$

where $\psi_1, \ldots, \psi_q$, $\rho_1, \ldots, \rho_p$ are real numbers such that $|\psi_j| < 1$, $j = 1, \ldots, q$, $|\rho_j| < 1$, $j = 1, \ldots, p$.

Let $\hat{\rho}_{j,ML}$, $\hat{\psi}_{j,ML}$ and $\hat{\sigma}^2_{ML}$ be the maximum likelihood estimators of ρ_j, ψ_j and σ^2, respectively. Since we have already evaluated I, J and K for the spectral density (2.4.10) (see (2.2.59), (2.2.61) and (2.2.63)), the equations (2.4.7) and (2.4.8) are easily solved. Then we have

Example 2.4.1. If $\theta = \rho_j$, then

$$g_0(\rho_j) = \frac{1}{2} \log \frac{1 + \rho_j}{1 - \rho_j}$$

and

$$c_0 = \frac{-\rho_j}{1 - \rho_j^2}.$$

That is

$$P_\theta^n\left[\sqrt{n(1-\rho_j^2)}\left\{\frac{1}{2}\log\frac{1+\hat\rho_{j,ML}}{1-\hat\rho_{j,ML}}-\frac{1}{2}\log\frac{1+\rho_j}{1-\rho_j}+\frac{\rho_j}{n(1-\rho_j^2)}\right\}\le x\right]=\Phi(x)+o(n^{-\frac{1}{2}}). \quad (2.4.11)$$

Example 2.4.2. If $\theta=\psi_j$, then

$$g_0(\psi_j)=\psi_j$$

and

$$c_0=\psi_j.$$

That is

$$P_\theta^n\left[\sqrt{\frac{n}{1-\psi_j^2}}(\hat\psi_{j,ML}-\psi_j-\frac{\psi_j}{n})\le x\right]=\Phi(x)+o(n^{-\frac{1}{2}}). \quad (2.4.12)$$

Example 2.4.3. If $\theta=\sigma^2$, then

$$g_0(\sigma^2)=\{\sigma^2\}^{\frac{1}{3}}$$

and

$$c_0=-\frac{2}{9}\sigma^{\frac{2}{3}}.$$

That is

$$P_\theta^n\left[\frac{3}{\sqrt2}\frac{\sqrt n}{\sigma^{\frac{2}{3}}}\left\{(\hat\sigma_{ML}^2)^{\frac{1}{3}}-(\sigma^2)^{\frac{1}{3}}+\frac{2\sigma^{\frac{2}{3}}}{9n}\right\}\le x\right]=\Phi(x)+o(n^{-\frac{1}{2}}). \quad (2.4.13)$$

We now turn to the transformation of the quasi-maximum likelihood estimator $\hat\theta_{qML}$ of θ, which maximizes the quasi-(log) likelihood defined by (2.2.43). Since we have already investigated the asymptotic properties of $\sqrt n(\hat\theta_{qML}-\theta)$ in (2.2.55) – (2.2.58), we can proceed as in the proof of Theorem 2.4.1 to develop the following.

Theorem 2.4.2. *Under Assumptions 2.2.1 - 2.2.4,*

$$P_\theta^n\left[\sqrt{nI(\theta)}\frac{g(\hat\theta_{qML})-g(\theta)-c/n}{g'(\theta)}\le x\right]$$
$$=\ \Phi(x)-\phi(x)\left[\frac{1}{\sqrt n}\left\{-\frac{B(\theta)}{I(\theta)^{\frac{1}{2}}}-\frac{K(\theta)}{6I(\theta)^{\frac{3}{2}}}-\frac{cI(\theta)^{\frac{1}{2}}}{g'(\theta)}\right\}\right.$$
$$\left.+\frac{1}{6\sqrt n}\left\{-\frac{3J(\theta)+2K(\theta)}{I(\theta)^{\frac{3}{2}}}+\frac{3g''(\theta)}{g'(\theta)I(\theta)^{\frac{1}{2}}}\right\}x^2\right]+o(n^{-\frac{1}{2}}). \quad (2.4.14)$$

Corollary 2.4.2. Under Assumptions 2.2.1 – 2.2.4, if $g_1(\theta)$ and c_1 satisfy

$$\frac{g_1''(\theta)}{g_1'(\theta)} = \frac{3J(\theta) + 2K(\theta)}{3I(\theta)} \tag{2.4.15}$$

and

$$c_1 = -g_1'(\theta)\left\{\frac{B(\theta)}{I(\theta)} + \frac{K(\theta)}{6I(\theta)^2}\right\}, \tag{2.4.16}$$

then

$$P_\theta^n\left[\sqrt{nI(\theta)}\frac{g_1(\hat{\theta}_{qML}) - g_1(\theta) - c_1/n}{g_1'(\theta)} \le x\right] = \Phi(x) + o(n^{-\frac{1}{2}}). \tag{2.4.17}$$

From (2.4.7) and (2.4.15) we can see $g_0(\theta) = g_1(\theta)$. Thus for the quasi-maximum likelihood estimators the same transformations as the maximum likelihood estimators give the normalizing transformation. We can also seek the constant c_1 which satisfies (2.4.16). Since evaluations of $B(\theta)$ are difficult for general ARMA(p, q) processes, we consider the ARMA$(1, 1)$ process with the spectral density

$$f_\theta(\lambda) = \frac{\sigma^2}{2\pi}\frac{|1 - \psi e^{i\lambda}|^2}{|1 - \rho e^{i\lambda}|^2}.$$

In this case $B(\psi)$, $B(\rho)$ and $B(\sigma^2)$ have been given in Section 2.2 (Cases 4 – 6), so we can give the explicit forms of c_1 for $\theta = \psi$, ρ and σ^2.

In multivariate analysis Konishi(1981) showed that Fisher's z-transformation $g_z(\theta) = \frac{1}{2}\log\{(1 + \theta)/(1 - \theta)\}$ and the transformation $\tilde{g}(\theta) = \theta^{1/3}$ are the normalizing transformations for the sample correlation coefficient and the latent roots of the sample covariance matrix respectively. Our examples show that $g_z(\theta)$ and $\tilde{g}(\theta)$ also are the normalizing transformations for the AR-part parameter ρ_j and the innovation variance σ^2 respectively. It may be noted that the AR-part parameter and the correlation coefficient represent a sort of correlation structure, and the innovation variance and the latent roots of the covariance matrix represent a sort of variance structure although our statistical models are essentially different from Konishi's ones.

Finally we give some numerical comparisons related to the approximation (2.4.11) in the first-order autoregressive process $X_t = \rho X_{t-1} + \epsilon_t$, where ϵ_t are i.i.d. $N(0, \sigma^2)$. Let

$$\tilde{\rho}_{ML} = (1 - n^{-1})\frac{\sum_{t=1}^{n-1} X_t X_{t+1}}{\sum_{t=2}^{n-1} X_t^2}.$$

It is known (Fujikoshi and Ochi(1984)) that

$$P_\rho^n\left[\sqrt{nI(\rho)}(\hat{\rho}_{ML} - \rho) \le x\right] - P_\rho^n\left[\sqrt{nI(\rho)}(\tilde{\rho}_{ML} - \rho) \le x\right] = o(n^{-1}),$$

where $\hat{\rho}_{ML}$ is the exact maximum likelihood estimator of ρ. Thus henceforth we use $\tilde{\rho}_{ML}$ in place of $\hat{\rho}_{ML}$.
Let

$$L(\rho, x) = \left|P_\rho^n\left[\sqrt{\frac{n}{1 - \rho^2}}(\tilde{\rho}_{ML} - \rho) \le x\right] - \Phi(x)\right|,$$

$$R(\rho, x) = \left|P_\rho^n\left[\sqrt{n(1 - \rho^2)}\left\{\frac{1}{2}\log\frac{1 + \tilde{\rho}_{ML}}{1 - \tilde{\rho}_{ML}} - \frac{1}{2}\log\frac{1 + \rho}{1 - \rho} + \frac{\rho}{n(1 - \rho^2)}\right\} \le x\right] - \Phi(x)\right|,$$

and

$$M(\rho, x) = \left| P_\rho^n \left[\sqrt{n(1 - \tilde{\rho}_{ML}^2)} \left\{ \frac{1}{2} \log \frac{1 + \tilde{\rho}_{ML}}{1 - \tilde{\rho}_{ML}} - \frac{1}{2} \log \frac{1 + \rho}{1 - \rho} + \frac{\tilde{\rho}_{ML}}{n(1 - \tilde{\rho}_{ML}^2)} \right\} \leq x \right] - \Phi(x) \right|.$$

Here the probability $P_\rho^n(\cdot)$ are computed by 5000 trials simulation. Table 2.4 in Appendix gives the values of $L(\rho, x)$, $R(\rho, x)$ and $M(\rho, x)$ for $n = 300$, $\rho = -0.9(0.3)0.9$ and $x = -2.0(0.5)2.0$. From the table we observe that the transformations proposed by us give better approximations than the usual normal approximations even if the normalizing factor is estimated.

2.5. Higher order asymptotic efficiency in time series regression models

In the previous sections we dealt with the processes with mean zero. In this section we discuss the higher order asymptotic theory for processes with non-zero mean. For an AR(1) model with unknown, constant mean Tanaka(1983) gave the third-order Edgeworth expansions of the distributions of the least squares estimators for the autoregressive coefficient and mean. Also Tanaka(1984) derived the second-order Edgeworth expansion of the joint distribution of the maximum likelihood estimator for an ARMA model with unknown mean. Furthermore we proceed to discuss more general models. Consider the statistical linear model

$$\mathbf{y} = X\beta + \mathbf{u}, \tag{2.5.1}$$

where $\mathbf{y}$ is an $n \times 1$ vector of observations, X is an $n \times p$ matrix of explanatory variables, β is a $p \times 1$ unknown vector, and $\mathbf{u}$ is a column vector of n disturbances. The error vector $\mathbf{u}$ is normally distributed with mean zero and nonsingular covariance matrix $V = V(\theta)$ where $\theta = (\theta_1, \ldots, \theta_q)'$ is an unknown parameter vector. If θ is known the best linear unbiased estimator of β is given by

$$\hat{\beta}(\theta) = [X'V(\theta)^{-1}X]^{-1}X'V(\theta)^{-1}\mathbf{y}.$$

If θ is unknown we can construct the generalized least squares estimator

$$\hat{\beta}(\hat{\theta}) = [X'V(\hat{\theta})^{-1}X]^{-1}X'V(\hat{\theta})^{-1}\mathbf{y},$$

where $\hat{\theta}$ is an appropriate consistent estimator of θ.
Rothenberg(1984) gave higher order approximations to the distribution of $\hat{\beta}(\hat{\theta})$ as follows.

Assumption 2.5.1.
(i) The estimator $\hat{\theta}$, when written as a function of $X\beta + \mathbf{u}$, does not depend on β and is an even function of $\mathbf{u}$.
(ii) The limiting distribution of $\sqrt{n}(\hat{\theta} - \theta)$ is $N(0, I(\theta)^{-1})$ where $I(\theta)$ is the asymptotic information matrix for θ.

Since it is convenient to work with scalars rather than vectors he considers the asymptotic distribution of $\mathbf{c}'\{\hat{\beta}(\hat{\theta}) - \beta\}$ where $\mathbf{c}$ is any constant p-dimensional vector. To obtain asymptotic approximations to its distribution, somewhat stronger assumptions are necessary.

Assumption 2.5.2. The standardized difference between $\mathbf{c}'\hat{\beta}(\hat{\theta})$ and $\mathbf{c}'\hat{\beta}(\theta)$ can be written as

$$\frac{\mathbf{c}'\{\hat{\beta}(\hat{\theta}) - \hat{\beta}(\theta)\}}{\sqrt{\mathbf{c}'(X'V(\theta)^{-1}X)^{-1}\mathbf{c}}} = \frac{1}{\sqrt{n}}Z_n + \frac{1}{n^2\sqrt{n}}R_n,$$

where Z_n possesses bounded moments as n tends to infinity; R_n is stochastically bounded with

$$P[|R_n| > (\log n)^r] = o(n^{-2}),$$

for some constant r.

Then the following theorem is proved.

Theorem 2.5.1. (Rothenberg(1984)). *Under Assumptions 2.5.1 and 2.5.2,*

$$P\left[\frac{\mathbf{c}'\{\hat{\beta}(\hat{\theta}) - \beta\}}{\sigma_n} \le x\right] = \Phi\left(x - \frac{x^3 - 3x}{24n^2}a_n\right) + o(n^{-2}), \qquad (2.5.2)$$

where $\sigma_n^2 = \mathbf{c}'(X'V(\theta)^{-1}X)^{-1}\mathbf{c}[1 + \frac{1}{n}Var(Z_n)]$, *and* $a_n = cum^{(4)}(Z_n)$.

More explicit expressions for a_n and σ_n^2 are needed. Let B be the $q \times q$ matrix having typical element

$$b_{ij} = \frac{\mathbf{c}'(X'V(\theta)^{-1}X)^{-1}X'V_i^{-1}MVM'V_j^{-1}X(X'V(\theta)^{-1}X)^{-1}\mathbf{c}}{\mathbf{c}'(X'V(\theta)^{-1}X)^{-1}\mathbf{c}},$$

where $M = I - X(X'V(\theta)^{-1}X)^{-1}X'V(\theta)^{-1}$ and $V_i^{-1} = \frac{\partial}{\partial \theta_i}V(\theta)^{-1}$.
Under some appropriate regularity conditions Rothenberg(1984) showed

$$\sigma_n^2 = \mathbf{c}'(X'V(\theta)^{-1}X)^{-1}\mathbf{c}\left[1 + \frac{1}{n}\text{tr}I(\theta)^{-1}B + O(n^{-2})\right],$$
$$a_n = 6\text{tr}I(\theta)^{-1}BI(\theta)^{-1}B + O(n^{-1}).$$

Although Rothenberg's formula (2.5.2) is very general it is for the linear combination $\mathbf{c}'\hat{\beta}(\hat{\theta})$, and its asymptotic optimality is not discussed. In this respect Maekawa(1985) dealt with the joint distribution of $\hat{\beta}(\hat{\theta})$, and discussed its third-order optimal property. He assumed the following in addition to those of Rothenberg.

Assumption 2.5.3. The distribution of $\begin{pmatrix} \sqrt{n}\{\hat{\beta}(\hat{\theta}) - \beta\} \\ \sqrt{n}(\hat{\theta} - \theta) \end{pmatrix}$ as $n \to \infty$ tends to the normal distribution $N\left[0, \begin{pmatrix} \Lambda_{11} & 0 \\ 0 & \Lambda_{22} \end{pmatrix}\right]$.

Then the stochastic expansion for $\sqrt{n}\{\hat{\beta}(\hat{\theta}) - \beta\}$ is derived in the form

$$\sqrt{n}\{\hat{\beta}(\hat{\theta}) - \beta\} = \mathbf{W}_1 + \frac{1}{\sqrt{n}}\mathbf{W}_2 + o_p(n^{-\frac{1}{2}}),$$

where $\mathbf{W}_i = O_p(1)$, $i = 1, 2$. Using this the following theorem is proved.

Theorem 2.5.2. (Maekawa(1985)).

(i) $\hat{\beta}(\hat{\theta})$ *belongs to the class* **D** *(see (2.3.61)),*

(ii) $\hat{\beta}(\hat{\theta})$ *is third-order asymptotically efficient in* **D** *(cf. Theorem 2.3.9).*

Now denote the tth component of (2.5.1) by

$$y_t = \mathbf{x}_t'\beta + u_t. \tag{2.5.3}$$

In view of the mean square error Toyooka(1985) derived the asymptotic expansions of $M_1 = E[\{\hat{\beta}(\hat{\theta}) - \beta\}\{\hat{\beta}(\hat{\theta}) - \beta\}']$ and $M_2 = E[\{\hat{\beta}(\theta) - \beta\}\{\hat{\beta}(\theta) - \beta\}']$ up to order $O(n^{-2})$ when u_t is an autoregressive process,

$$u_t = \theta u_{t-1} + \epsilon_t, \quad |\theta| < 1,$$

where $\{\epsilon_t\}$ is a sequence of independently identically normally distributed random variables with mean zero and variance σ^2, and $\hat{\theta}$ is an appropriate estimator of θ. He also gave a sufficient condition for $M_1 - M_2 = o(n^{-2})$.

Furthermore Toyooka(1986) extended the above results to the case when the process $\{u_t\}$ in (2.5.3) is a general linear process,

$$u_t = \sum_{j=0}^{\infty} g_j(\theta)\epsilon_{t-j}, \tag{2.5.4}$$

with $g_0(\theta) = 1$, $\sum_{j=0}^{\infty} g_j(\theta)^2 < \infty$, and $\{\epsilon_t\}$ is a sequence of i.i.d. $N(0, \sigma^2)$ random variables. His discussion is as follows. Let

$$U_n(\theta) = \frac{1}{2\pi} \int_{-\pi}^{\pi} \left|\sum_{t=1}^{n} \tilde{u}_t e^{i\lambda t}\right|^2 \cdot \left|\sum_{j=0}^{\infty} g_j(\theta)e^{i\lambda j}\right|^{-2} d\lambda$$

where $\tilde{u}_t = y_t - \mathbf{x}_t' \cdot \hat{\beta}$, with $\hat{\beta} = (X'X)^{-1}X'\mathbf{y}$. The estimator $\hat{\theta}$ of θ is defined by the value of θ which minimizes $U_n(\theta)$ with respect to θ (see Walker(1964)). The following regularity conditions on the regression functions $\{\mathbf{x}_t = (x_{1t}, \ldots, x_{pt})'\}$ are imposed.

X1. $d_i^2 = \sum_{t=1}^{n} x_{it}^2 \to \infty$ as $n \to \infty$ for $i = 1, \ldots, p$.

X2. $\lim_{n\to\infty} x_{i,n+1}^2/d_i^2 = 0$ for $i = 1, \ldots, p$.

X3. The limit

$$\lim_{n\to\infty} \frac{1}{n}\sum_{t=1}^{n-h} \mathbf{x}_t \mathbf{x}_{t+h}' = R(h)$$

exists, and there exists a matrix-valued regression spectral measure $M_x(\lambda)$ such that

$$R(h) = \int_{-\pi}^{\pi} e^{ih\lambda} dM_x(\lambda).$$

X4. $R(0)$ is nonsingular.

Denote the (t, i)-th element of $Z = V(\theta)^{-1}X$ by z_{it}. We set new assumptions Z1 – Z4 for z_{it} which are X1 – X4 with replacing x_{it} by z_{it}, respectively. Here the regression spectral matrix of $\{z_{it}\}$ is written as $M_z(\lambda)$. Then,

Theorem 2.5.3. (Toyooka(1986)). *Under X1 - X4 and Z1 - Z4,*

$$E[\{\hat{\boldsymbol{\beta}}(\hat{\theta}) - \hat{\boldsymbol{\beta}}(\theta)\}\{\hat{\boldsymbol{\beta}}(\hat{\theta}) - \hat{\boldsymbol{\beta}}(\theta)\}'] = \frac{1}{n^2}\nu + o(n^{-2}), \tag{2.5.5}$$

where ν is explicitly given in terms of the spectral density of $\{u_t\}$ and the regression spectrum $M_x(\lambda)$ and $M_z(\lambda)$.

Theorem 2.5.4. (Toyooka(1986)). *Under X1 - X4 and Z1 - Z4, if $M_x(\lambda)$ and $M_z(\lambda)$ each have only one jump point at $\lambda = 0$, then the coefficient ν in (2.5.5) vanishes, and*

$$E[\{\hat{\boldsymbol{\beta}}(\hat{\theta}) - \boldsymbol{\beta}\}\{\hat{\boldsymbol{\beta}}(\hat{\theta}) - \boldsymbol{\beta}\}'] - E[\{\hat{\boldsymbol{\beta}}(\theta) - \boldsymbol{\beta}\}\{\hat{\boldsymbol{\beta}}(\theta) - \boldsymbol{\beta}\}'] = o(n^{-2}).$$

We now turn to the third-order asymptotic properties of $\hat{\boldsymbol{\beta}}(\hat{\theta})$ and the least squares estimator in terms of the probability concentration around the true value. To avoid unnecessary complexity of vectors we confine ourselves to the regression model (2.5.1) with $p = 1$, i.e.,

$$\mathbf{y} = \mathbf{x} \cdot \beta + \mathbf{u}, \tag{2.5.6}$$

where $\mathbf{x} = (x_1, \ldots, x_n)'$ and β represent X and $\boldsymbol{\beta}$ with $p = 1$, respectively. Here we set down

Assumption 2.5.4.

$$\frac{1}{n}\mathbf{x}'V(\boldsymbol{\theta})^{-1}\mathbf{x} = A + \frac{1}{n}B + o(n^{-1}).$$

First, we derive the third-order bound distribution for the third-order AMU estimators of β. The likelihood function based on $\mathbf{y}$ is given by

$$L(\beta) = (2\pi)^{-\frac{n}{2}}|V(\boldsymbol{\theta})|^{-\frac{1}{2}}\exp{-\frac{1}{2}\{\mathbf{y} - \mathbf{x}\beta\}'V(\boldsymbol{\theta})^{-1}\{\mathbf{y} - \mathbf{x}\beta\}}.$$

Consider the problem of testing hypothesis $H : \beta = \beta_0 + t/\sqrt{n}$ $(t > 0)$ against alternative $K : \beta = \beta_0$. Let

$$LR = \log\frac{L(\beta_0)}{L(\beta_1)},$$

where $\beta_1 = \beta_0 + t/\sqrt{n}$. Then

$$LR = -\frac{t}{\sqrt{n}}\mathbf{x}'V(\boldsymbol{\theta})^{-1}(\mathbf{y} - \mathbf{x}\beta_0) + \frac{t^2}{2n}\mathbf{x}'V(\boldsymbol{\theta})^{-1}\mathbf{x}.$$

From Assumption 2.5.4 and Gaussianity of $\mathbf{y}$, we can evaluate the asymptotic cumulants of LR as follows;

$$E_{\beta_0}\{LR\} = \frac{t^2}{2}(A + \frac{1}{n}B) + o(n^{-1}), \tag{2.5.7}$$

$$\text{cum}_{\beta_0}\{LR, LR\} = t^2(A + \frac{1}{n}B) + o(n^{-1}), \tag{2.5.8}$$

$$\text{cum}_{\beta_0}^{(J)}\{LR\} = 0, \quad \text{for} \quad J \geq 3. \tag{2.5.9}$$

Similarly,

$$E_{\beta_1}\{LR\} = -\frac{t}{\sqrt{n}}\mathbf{x}'V(\boldsymbol{\theta})^{-1}\mathbf{x}\left(\frac{t}{\sqrt{n}}\right) + \frac{t^2}{2n}\mathbf{x}'V(\boldsymbol{\theta})^{-1}\mathbf{x}$$

$$= -\frac{t^2}{2}(A + \frac{1}{n}B) + o(n^{-1}), \tag{2.5.10}$$

$$\mathrm{cum}_{\beta_1}\{LR, LR\} = t^2(A + \frac{1}{n}B) + o(n^{-1}), \tag{2.5.11}$$

$$\mathrm{cum}_{\beta_1}^{(J)}\{LR\} = 0, \quad \text{for} \quad J \geq 3. \tag{2.5.12}$$

In view of (2.5.10) – (2.5.12) and (2.1.15), if we put $U_n = LR + \frac{t^2}{2}(A + \frac{B}{n})$, then we can see

$$P_{\beta_1}^n[U_n \leq 0] = \frac{1}{2} + o(n^{-1}),$$

$$P_{\beta_1}^n[U_n \geq 0] = \frac{1}{2} + o(n^{-1}),$$

which implies U_n is third-order AMU. From (2.5.7) – (2.5.9) and (2.1.15), we have the bound distribution for third-order AMU estimators;

$$P_{\beta_0}^n\left[-U_n + t^2(A + \frac{B}{n}) \leq t^2(A + \frac{B}{n})\right] = \Phi\left\{t\sqrt{A}\right\} + \frac{Bt}{2\sqrt{A}n}\phi\left\{t\sqrt{A}\right\} + o(n^{-1}). \tag{2.5.13}$$

For $T_n = \sqrt{n}\{\hat{\beta}(\theta) - \beta_0\}$ we have

$$E_{\beta_0}\{T_n\} = 0,$$

$$\mathrm{cum}_{\beta_0}\{T_n, T_n\} = A^{-1} - \frac{1}{n}A^{-2}B + o(n^{-1}),$$

$$\mathrm{cum}_{\beta_0}^{(J)}\{T_n\} = 0, \quad \text{for} \quad J \geq 3,$$

which establish

Theorem 2.5.5. *Under Assumption 2.5.4,*

$$P_{\beta_0}^n\left[\sqrt{n}\{\hat{\beta}(\boldsymbol{\theta}) - \beta_0\} \leq t\right] = \Phi\left\{t\sqrt{A}\right\} + \frac{Bt}{2\sqrt{A}n}\phi\left\{t\sqrt{A}\right\} + o(n^{-1}).$$

That is, $\hat{\beta}(\boldsymbol{\theta})$ is third-order asymptotically efficient in the class $\mathbf{A}_3$ of third-order AMU estimators.

We next consider the least squares estimator $\hat{\beta}_{LS} = (\mathbf{x}'\mathbf{x})^{-1}\mathbf{x}'\mathbf{y}$. The variance is

$$E_{\beta_0}(\hat{\beta}_{LS} - \beta_0)^2 = (\mathbf{x}'\mathbf{x})^{-1}\mathbf{x}'V(\boldsymbol{\theta})\mathbf{x}(\mathbf{x}'\mathbf{x})^{-1}.$$

Here we impose the following assumption.

Assumption 2.5.5.
(i) $\mathbf{x}'\mathbf{x} = O(n)$.
(ii) $n(\mathbf{x}'\mathbf{x})^{-1}\mathbf{x}'V(\boldsymbol{\theta})\mathbf{x}(\mathbf{x}'\mathbf{x})^{-1} = A^{-1} + \frac{C}{n} + o(n^{-1})$.

The above assumption is natural if we assume Grenander's conditions X1 – X4 for the regression functions. From Schwarz's inequality we have

$$(\mathbf{x}'\mathbf{x})^2 = \left\{\mathbf{x}'V(\boldsymbol{\theta})^{\frac{1}{2}}V(\boldsymbol{\theta})^{-\frac{1}{2}}\mathbf{x}\right\}^2$$
$$\leq \{\mathbf{x}'V(\boldsymbol{\theta})\mathbf{x}\}\{\mathbf{x}'V(\boldsymbol{\theta})^{-1}\mathbf{x}\};$$

that is,

$$(\mathbf{x}'\mathbf{x})^{-1}\mathbf{x}'V(\boldsymbol{\theta})\mathbf{x}(\mathbf{x}'\mathbf{x})^{-1} \geq \{\mathbf{x}'V(\boldsymbol{\theta})^{-1}\mathbf{x}\}^{-1}. \tag{2.5.14}$$

It follows from this and Assumptions 2.5.4 and 2.5.5 that

$$A^{-1} + \frac{C}{n} + o(n^{-1}) \geq \left\{A + \frac{B}{n} + o(n^{-1})\right\}^{-1}, \quad \text{for all} \quad n,$$

which implies

$$C + A^{-2}B \geq 0. \tag{2.5.15}$$

Putting $W_n = \sqrt{n}\{\hat{\beta}_{LS} - \beta_0\}$ we have

$$E_{\beta_0}\{W_n\} = 0,$$

$$\text{cum}_{\beta_0}\{W_n, W_n\} = A^{-1} + \frac{C}{n} + o(n^{-1}),$$

$$\text{cum}_{\beta_0}^{(J)}\{W_n\} = 0, \quad \text{for} \quad J \geq 3.$$

Thus the Edgeworth expansion of W_n is

$$P_{\beta_0}^n\left[\sqrt{n}(\hat{\beta}_{LS} - \beta_0) \leq t\right] = \Phi\left\{t\sqrt{A}\right\} - \frac{A^{3/2}C}{2n}t\phi\left\{t\sqrt{A}\right\} + o(n^{-1}). \tag{2.5.16}$$

Summarizing the above we have

Theorem 2.5.6. *Under Assumptions 2.5.4 and 2.5.5,*

$$\lim_{n\to\infty} n\left[P_{\beta_0}^n\left\{\sqrt{n}(\hat{\beta}(\theta) - \beta_0) \leq t\right\} - P_{\beta_0}^n\left\{\sqrt{n}(\hat{\beta}_{LS} - \beta_0) \leq t\right\}\right]$$
$$= \frac{B + A^2C}{2\sqrt{A}}t\phi\left\{t\sqrt{A}\right\} \geq 0, \quad \text{for} \quad t \geq 0.$$

That is, i) $\hat{\beta}_{LS}$ is always second-order asymptotically efficient in $\mathbf{A}_2$. *ii) $\hat{\beta}_{LS}$ is third-order asymptotically efficient in* $\mathbf{A}_3$ *if and only if* $B + A^2C = 0$.

To obtain the explicit forms of A, B and C, henceforth, we specify our regression model $\mathbf{y} = \mathbf{x} \cdot \beta + \mathbf{u}$, where $\mathbf{x} = (x_1, \ldots, x_n)'$ and $\mathbf{u} = (u_1, \ldots, u_n)'$, as follows. We assume that u_t is generated by the relation

$$u_t - \alpha u_{t-1} = \epsilon_t - \rho\epsilon_{t-1}, \quad |\alpha|, |\rho| < 1, \tag{2.5.17}$$

where $\{\epsilon_t\}$ are mutually independent normal random variables each with mean zero and variance σ^2. As for $\mathbf{x} = (x_1, \ldots, x_n)'$ we confine ourselves to the following two cases;

Case 1. $x_t = 1$, $t = 1, \ldots, n$.

Case 2. $x_t = \cos \lambda t$, $\lambda = 2\pi j/n$ $(j;$ integer$)$, $t = 1, \ldots, n$.

For the ARMA(1,1) model defined by (2.5.17), Galbraith and Galbraith(1974) gave the exact expression of the inverse matrix of $V(\theta) = E\mathbf{u}\mathbf{u}'$ (see Section 2.3). Using their exact expression we can evaluate A, B and C;

Case 1.

$$A = \frac{(1-\alpha)^2}{\sigma^2(1-\rho)^2}, \quad B = -\frac{2(\rho-\alpha)(1-\alpha)}{\sigma^2(1-\rho)^3}, \quad C = -\frac{2\sigma^2(1-\alpha\rho)(\alpha-\rho)}{(1+\alpha)(1-\alpha)^3}$$

Case 2.

$$A = \frac{|1-\alpha e^{i\lambda}|^2}{2\sigma^2|1-\rho e^{i\lambda}|^2}, \quad B = \frac{(\rho-\alpha)\{(\alpha+\rho)\cos^2\lambda - 2(1+\alpha\rho)\cos\lambda + \alpha + \rho\}}{\sigma^2|1-\rho e^{i\lambda}|^4},$$

$$C = \frac{8\sigma^2(1-\alpha\rho)(\alpha-\rho)(\alpha\cos\lambda - 1)(\cos\lambda - \alpha)}{(1-\alpha^2)|1-\alpha e^{i\lambda}|^4}.$$

It is easy to show that $B + A^2 C = 0$ if and only if $\alpha = \rho$. Therefore $\hat{\beta}_{LS}$ is not third-order asymptotically efficient except for the case when the residual process is white. For the residual process (2.5.17) with $\theta = (\alpha, \rho, \sigma^2)'$ we have the following theorems.

Theorem 2.5.7. *In Case 1 we have*

$$nE_{\beta_0}[\hat{\beta}_{LS} - \beta_0]^2 = \frac{\sigma^2(1-\rho)^2}{(1-\alpha)^2} - \frac{2\sigma^2}{n}\frac{(1-\alpha\rho)(\alpha-\rho)}{(1+\alpha)(1-\alpha)^3} + o(n^{-1}), \qquad (2.5.18)$$

$$nE_{\beta_0}[\hat{\beta}(\theta) - \beta_0]^2 = \frac{\sigma^2(1-\rho)^2}{(1-\alpha)^2} + \frac{2\sigma^2}{n}\frac{(\rho-\alpha)(1-\rho)}{(1-\alpha)^3} + o(n^{-1}), \qquad (2.5.19)$$

$$nE_{\beta_0}[\hat{\beta}_{LS} - \beta_0]^2 - nE_{\beta_0}[\hat{\beta}(\theta) - \beta_0]^2$$
$$= \frac{2\sigma^2}{n}\frac{(\rho-\alpha)^2}{(1-\alpha)^3(1+\alpha)} + o(n^{-1}), \qquad (2.5.20)$$
$$= D_1(\alpha, \rho, n) + o(n^{-1}), \quad (say).$$

Theorem 2.5.8. *In Case 2 we have*

$$nE_{\beta_0}[\hat{\beta}_{LS} - \beta_0]^2$$
$$= 2\sigma^2\frac{|1-\rho e^{i\lambda}|^2}{|1-\alpha e^{i\lambda}|^2} + \frac{8\sigma^2}{n}\frac{(1-\alpha\rho)(\alpha-\rho)(\alpha\cos\lambda - 1)(\cos\lambda - \alpha)}{(1-\alpha^2)|1-\alpha e^{i\lambda}|^4} + o(n^{-1}), \quad (2.5.21)$$

$$nE_{\beta_0}[\hat{\beta}(\theta) - \beta_0]^2$$
$$= 2\sigma^2\frac{|1-\rho e^{i\lambda}|^2}{|1-\alpha e^{i\lambda}|^2}$$
$$- \frac{4\sigma^2(\rho-\alpha)}{n}\frac{\{(\alpha+\rho)\cos^2\lambda - 2(1+\alpha\rho)\cos\lambda + \alpha + \rho\}}{|1-\alpha e^{i\lambda}|^4} + o(n^{-1}), \quad (2.5.22)$$

$$
nE_{\beta_0}[\hat{\beta}_{LS} - \beta_0]^2 - nE_{\beta_0}[\hat{\beta}(\theta) - \beta_0]^2
$$
$$
= \frac{4\sigma^2(\alpha - \rho)^2\{(\alpha^2 + 1)\cos^2\lambda - 4\alpha\cos\lambda + \alpha^2 + 1\}}{n|1 - \alpha e^{i\lambda}|^4(1 - \alpha^2)} + o(n^{-1}) \qquad (2.5.23)
$$
$$
= D_2(\alpha, \rho, \lambda, n) + o(n^{-1}), \quad (say).
$$

In the above theorems we evaluated the third-order differences $D_1(\alpha, \rho, n)$ and $D_2(\alpha, \rho, \lambda, n)$ of the mean square errors of $\hat{\beta}_{LS}$ and $\hat{\beta}(\theta)$. Table 2.5.1 in Appendix gives the values of $D_1(\alpha, \rho, n)$ for $\sigma^2 = 1$, $n = 30$, $\alpha = -0.875(0.125)0.875$ and $\rho = -0.875(0.125)0.875$. Tables 2.5.2 – 2.5.4 in Appendix give the values of $D_2(\alpha, \rho, \pi/6, n)$, $D_2(\alpha, \rho, \pi/2, n)$ and $D_2(\alpha, \rho, 5\pi/6, n)$ for $\sigma^2 = 1$, $n = 30$, $\alpha = -0.875(0.125)0.875$ and $\rho = -0.875(0.125)0.875$, respectively. From these tables we observe that the difference between the mean square errors of $\hat{\beta}_{LS}$ and $\hat{\beta}(\theta)$ becomes larger as $\{u_t\}$ tends to an ARIMA process ($|\alpha| \to 1$).

Finally we investigate the third-order asymptotic properties of $\hat{\beta}(\hat{\theta})$ where $\hat{\theta}$ is the quasi-maximum likelihood estimator of $\theta = (\alpha, \rho, \sigma^2)'$ which is defined by (2.3.62). In Case 1 (i.e., $\mathbf{x} = (1, \ldots, 1)'$), denote the tth element of $\mathbf{z} = V(\theta)^{-1}\mathbf{x}$ by z_t. Using the explicit expression of $V(\theta)^{-1}$ for (2.5.17) (see Section 2.3 or Galbraith and Galbraith(1974)) it is easy to check that $\{z_t\}$ satisfies Toyooka's conditions Z1 – Z4. Also we can show that the regression spectrum of $\mathbf{x}$ and $\mathbf{z}$ in Case 1 both have only one jump point at $\lambda = 0$. Thus Theorems 2.5.4 and 2.5.5 together yield

Theorem 2.5.9 *In Case 1, $\hat{\beta}(\hat{\theta})$ is third-order asymptotically efficient in* $\mathbf{A}_3$.

CHAPTER 3

VALIDITY OF EDGEWORTH EXPANSIONS IN TIME SERIES ANALYSIS

In Chapter 2 the formal Edgeworth expansions were used for the studies of higher order asymptotic theory. Thus the proofs of their validities have been required. Götze and Hipp(1983) showed that formal Edgeworth expansions are valid for sums of weakly dependent vectors. Durbin(1980a) and Taniguchi(1984) showed the validity of Edgeworth expansions of statistics derived from observations which are not necessarily independent and identically distributed. However their sufficient conditions for the validity are hard to check even in the fundamental statistics. In this chapter we first give Berry-Esseen theorems for quadratic forms of a Gaussian stationary process. Second, we propose a generalized maximum likelihood estimator which includes the maximum likelihood estimator and the quasi-maximum likelihood estimator as special cases. Consider a Gaussian ARMA process with unknown parameter θ, and let $\hat{\theta}_n$ be the generalized maximum likelihood estimator of θ. Then we derive the Edgeworth expansion of the distribution of $\hat{\theta}_n$ up to third order, and prove its validity.

3.1. Berry-Esseen theorems for quadratic forms of Gaussian stationary processes

The classical Berry-Esseen theorem for the arithmetic sum of independent random variables has developed in various directions. For example Does(1982) gave Berry-Esseen bounds for simple linear rank statistics from independent and identically distributed random variables. A similar result for linear combinations of order statistics was given in Helmers(1981). For stochastic processes Erickson(1974) gave L_1-Berry-Esseen bounds for sums of m-dependent random variables.

In this section we shall give a Berry-Esseen theorem for quadratic forms of a Gaussian stationary process $\{X_t\}$. Put

$$Z_j = \frac{1}{\sqrt{n}}\left\{\sum_{l=1}^{n-j+1}(X_l X_{l+j-1} - EX_l X_{l+j-1})\right\}, \quad j = 1, \ldots, p.$$

Let $\Phi_V(\cdot)$ be the normal distribution with covariance matrix V and mean 0. Then we shall prove that

$$\sup_{C \in \Xi}|P\{(Z_1, \ldots, Z_p) \in C\} - \Phi_V(C)| = O(n^{-\frac{1}{2}}),$$

where Ξ is the class of all Borel-measurable convex subsets of R^p. The quadratic forms are fundamental and important quantities in time series analysis (e.g., the Yule-Walker equations are described by them).

Throughout this section we assume that $\{X_t; t = 0, \pm 1, \pm 2, \ldots\}$ is a Gaussian stationary process with spectral density $f(\lambda)$ and mean 0, and that $f(\lambda)$ belongs to $\mathbf{D}_1$ defined in (2.2.1). Suppose that a stretch $\mathbf{X}_n = (X_1, \ldots, X_n)'$ of the series $\{X_t\}$ is available. Denote the covariance matrix of $\mathbf{X}_n$ by Σ_n.

Using the $n \times n$ matrix

$$L = \begin{pmatrix} 0 & & & 0 \\ 1 & 0 & & \\ & \ddots & \ddots & \\ 0 & & 1 & 0 \end{pmatrix},$$

we can rewrite Z_j as

$$Z_j = \frac{1}{2\sqrt{n}}\{\mathbf{X}_n' G_{j-1}\mathbf{X}_n - \mathrm{tr}G_{j-1}\Sigma_n\},$$

where $G_j = L^j + L'^j$, $L^0 = I_n$, $G_0 = 2I_n$, and I_n is the $n \times n$ identity matrix.

Next we derive asymptotic expansions of the characteristic function $\Psi(\mathbf{t})$ of $\mathbf{Z} = (Z_1, \ldots, Z_p)'$. Direct computation yields

$$\begin{aligned} \Psi(\mathbf{t}) &= Ee^{i\mathbf{t}'\mathbf{Z}} \\ &= \det\left\{I_n - \frac{i}{\sqrt{n}}\Sigma_n(t_1 G_0 + t_2 G_1 + \cdots + t_p G_{p-1})\right\}^{-\frac{1}{2}} \\ &\quad \times \exp -\frac{i}{2\sqrt{n}}(t_1\mathrm{tr}G_0\Sigma_n + t_2\mathrm{tr}G_1\Sigma_n + \cdots + t_p\mathrm{tr}G_{p-1}\Sigma_n), \end{aligned} \tag{3.1.1}$$

where $\mathbf{t} = (t_1, \ldots, t_p)'$. Thus we have

$$\begin{aligned} \log\Psi(\mathbf{t}) &= -\frac{1}{2}\sum_{j=1}^n \log(1 - \frac{i}{\sqrt{n}}\rho_j) \\ &\quad -\frac{i}{2\sqrt{n}}(t_1\mathrm{tr}G_0\Sigma_n + t_2\mathrm{tr}G_1\Sigma_n + \cdots + t_p\mathrm{tr}G_{p-1}\Sigma_n), \end{aligned} \tag{3.1.2}$$

where ρ_j is the jth latent root of $\Sigma_n^{1/2}(t_1 G_0 + \cdots + t_p G_{p-1})\Sigma_n^{1/2}$. Of course each ρ_j is a real number. Using Taylor's formula (e.g., Bhattacharya and Rao(1976, p.57)), we have

$$\log(1 - ih) = -ih + \frac{h^2}{2} + \frac{ih^3}{3} + \frac{h^4}{3!}\int_0^1 (1-v)^3 \frac{-6}{(1-ivh)^4}\,dv, \quad \text{for} \quad h \in R^1. \tag{3.1.3}$$

Notice that

$$\left|\int_0^1 (1-v)^3 \frac{-6}{(1-ivh)^4}\,dv\right| \le 6. \tag{3.1.4}$$

From (3.1.3) and (3.1.4),

$$\begin{aligned} \log\Psi(\mathbf{t}) &= -\frac{1}{2}\sum_{j=1}^n \left\{-\frac{i\rho_j}{\sqrt{n}} + \frac{\rho_j^2}{2n} + \frac{i\rho_j^3}{3n\sqrt{n}} + R_4(j)\right\} \\ &\quad -\frac{i}{2\sqrt{n}}(t_1\mathrm{tr}\Sigma_n G_0 + \cdots + t_p\mathrm{tr}\Sigma_n G_{p-1}), \end{aligned} \tag{3.1.5}$$

where

$$|R_4(j)| \le \left(\frac{\rho_j}{\sqrt{n}}\right)^4.$$

Thus

$$\log \Psi(\mathbf{t}) = -\frac{1}{4n}\mathrm{tr}\{\Sigma_n(t_1G_0 + t_2G_1 + \cdots + t_pG_{p-1})\}^2$$

$$-\frac{i}{6n\sqrt{n}}\mathrm{tr}\{\Sigma_n(t_1G_0 + t_2G_1 + \cdots + t_pG_{p-1})\}^3 + R_4, \qquad (3.1.6)$$

where

$$|R_4| \le \frac{1}{2n^2}\mathrm{tr}\{\Sigma_n(t_1G_0 + t_2G_1 + \cdots + t_pG_{p-1})\}^4$$

After slight modification for Theorem 2.2.1 we get the following lemma.

Lemma 3.1.1. *Suppose that* $f_1(\lambda), \ldots, f_s(\lambda) \in \mathbf{D}_1$.
We define $\Gamma_1, \ldots, \Gamma_s$, *the* $n \times n$ *Toeplitz type matrices, by*

$$\Gamma_j = \left(\int_{-\pi}^{\pi} e^{i(m-l)\lambda} f_j(\lambda)\, d\lambda\right), \quad m, l = 1, \ldots, n.$$

Then

$$\frac{1}{n}\mathrm{tr}\Gamma_1 \ldots \Gamma_s = (2\pi)^{s-1} \int_{-\pi}^{\pi} \prod_{j=1}^{s} f_j(\lambda)\, d\lambda + O(n^{-1}).$$

Notice that G_j, $j = 0, \ldots, p-1$, have the following Toeplitz forms;

$$G_j = \left(\int_{-\pi}^{\pi} \frac{\cos j\lambda}{\pi} e^{i(m-l)\lambda}\, d\lambda\right), \quad m, l = 1, \ldots, n, \qquad (3.1.7)$$

$j = 0, \ldots, p-1$. We set down

$$\tilde{V}_{jk} = \frac{1}{2n}\mathrm{tr}\Sigma_n G_{j-1}\Sigma_n G_{k-1},$$

$$\tilde{W}_{jkm} = \frac{1}{n}\mathrm{tr}\Sigma_n G_{j-1}\Sigma_n G_{k-1}\Sigma_n G_{m-1},$$

$$\tilde{R}_{jkml} = \frac{1}{n}\mathrm{tr}\Sigma_n G_{j-1}\Sigma_n G_{k-1}\Sigma_n G_{m-1}\Sigma_n G_{l-1},$$

$$j, k, m, l = 1, \ldots, p.$$

Then using Lemma 3.1.1 and (3.1.7) we get;

$$\tilde{V}_{jk} = V_{jk} + O(n^{-1}),$$

$$\tilde{W}_{jkm} = W_{jkm} + O(n^{-1}),$$

$$\tilde{R}_{jkml} = R_{jkml} + O(n^{-1}),$$

where

$$V_{jk} = 4\pi \int_{-\pi}^{\pi} f(\lambda)^2 \cos(j-1)\lambda \cdot \cos(k-1)\lambda\, d\lambda,$$

$$W_{jkm} = 32\pi^2 \int_{-\pi}^{\pi} f(\lambda)^3 \cos(j-1)\lambda \cdot \cos(k-1)\lambda \cdot \cos(m-1)\lambda\, d\lambda,$$

$$R_{jkml} = 2^7\pi^3 \int_{-\pi}^{\pi} f(\lambda)^4 \cos(j-1)\lambda \cdot \cos(k-1)\lambda \cdot \cos(m-1)\lambda \cdot \cos(l-1)\lambda\, d\lambda,$$

$$j, k, m, l = 1, \ldots, p.$$

Let V be the $p \times p$ matrix whose (j,k)-th element is V_{jk}. Since $f(\lambda)$ belongs to $\mathbf{D}_1$ and is continuous on $[-\pi, \pi]$ there exists $\delta > 0$ such that $f(\lambda) \geq \delta$ for all $\lambda \in [-\pi, \pi]$. Thus we have

$$
\begin{aligned}
\mathbf{b}'V\mathbf{b} &= 4\pi \int_{-\pi}^{\pi} f(\lambda)^2 \left| \sum_{j=1}^{p} b_j \cos(j-1)\lambda \right|^2 d\lambda \\
&\geq 4\pi\delta^2 \int_{-\pi}^{\pi} \left| \sum_{j=1}^{p} b_j \cos(j-1)\lambda \right|^2 d\lambda \\
&= 4\pi^2\delta^2 (2b_1^2 + b_2^2 + \cdots + b_p^2) \geq 4\pi^2\delta^2 > 0,
\end{aligned}
$$

for any $\mathbf{b} = (b_1, \ldots, b_p)'$, $\mathbf{b}'\mathbf{b} = 1$. Thus V is a positive definite matrix. We set down

$$
W = \max_{j,k,m} |W_{jkm}|,
$$

$$
R = \max_{j,k,m,l} |R_{jkml}|
$$

and

$$
\|\mathbf{t}\| = (t_1^2 + \cdots + t_p^2)^{\frac{1}{2}}.
$$

Let v the smallest latent root of V. We choose a positive number v_0 so that $0 < v_0 < v$.

Lemma 3.1.2. *Assume that a positive constant d_1 satisfies*

$$
0 < d_1 < \frac{\sqrt{W^2 p^3 + 36 R p^2 v_0} - W p^{3/2}}{6 R p^2}. \tag{3.1.8}
$$

If we take n sufficiently large, then for all $\mathbf{t}$ satisfying $\|\mathbf{t}\| \leq d_1 \sqrt{n}$, the following relation holds

$$
\begin{aligned}
&\left| \Psi(\mathbf{t}) - \exp\left\{-\frac{1}{2}\mathbf{t}'V\mathbf{t}\right\} \left\{ 1 + \frac{i^3}{6\sqrt{n}} \sum_j \sum_k \sum_m W_{jkm} t_j t_k t_m \right\} \right| \\
&\qquad \leq \exp\left\{-\frac{1}{2}\mathbf{t}'Q_1\mathbf{t}\right\} \times \|\mathbf{t}\|^6 O(n^{-1}) \\
&\qquad\quad + \exp\left\{-\frac{1}{2}\mathbf{t}'V\mathbf{t}\right\} \{\|\mathbf{t}\|^2 O(n^{-1}) + \|\mathbf{t}\|^4 O(n^{-1})\},
\end{aligned} \tag{3.1.9}
$$

where Q_1 is a positive definite matrix.

Proof. By (3.1.6) we have

$$
\log \Psi(\mathbf{t}) + \frac{1}{2} \sum_j \sum_k \tilde{V}_{jk} t_j t_k + \frac{i}{6\sqrt{n}} \sum_j \sum_k \sum_m \tilde{W}_{jkm} t_j t_k t_m = R_4, \tag{3.1.10}
$$

where

$$
|R_4| \leq \frac{1}{2n} \sum_j \sum_k \sum_m \sum_l \tilde{R}_{jkml} t_j t_k t_m t_l.
$$

Thus

$$\left|\log \Psi(t) + \frac{1}{2}\sum_j \sum_k V_{jk} t_j t_k + \frac{i}{6\sqrt{n}}\sum_j \sum_k \sum_m W_{jkm} t_j t_k t_m\right|$$

$$\leq \|t\|^2 O(n^{-1}) + \|t\|^3 O(n^{-\frac{3}{2}}) + \frac{Rp^2}{2n}\|t\|^4 + \|t\|^4 O(n^{-2}). \tag{3.1.11}$$

While from (3.1.10) we have

$$\left|\log \Psi(t) + \frac{1}{2}t'Vt\right|$$

$$\leq \frac{Wp^{3/2}}{6\sqrt{n}}\|t\|^3 + \|t\|^3 O(n^{-\frac{3}{2}}) + \|t\|^2 O(n^{-1}) + \frac{Rp^2}{2n}\|t\|^4 + \|t\|^4 O(n^{-2})$$

$$\leq \left\{\left(\frac{Wp^{3/2}d_1}{6} + \frac{Rp^2 d_1^2}{2}\right) + O(n^{-1})\right\}\|t\|^2, \quad \text{for } \|t\| \leq d_1\sqrt{n} \tag{3.1.12}$$

Use of (3.1.11), (3.1.12) and relation

$$|e^x - 1 - x| \leq \frac{|x|^2}{2}e^{|x|}$$

yields

$$\left|\Psi(t) - \exp\left\{-\frac{1}{2}t'Vt\right\}\left\{1 + \frac{i^3}{6\sqrt{n}}\sum_j \sum_k \sum_m W_{jkm} t_j t_k t_m\right\}\right|$$

$$\leq \exp\left\{-\frac{1}{2}t'Vt\right\} \times \left|\exp\left\{\log \Psi(t) + \frac{1}{2}t'Vt\right\} - \left(1 + \log \Psi(t) + \frac{1}{2}t'Vt\right)\right|$$

$$+ \exp\left\{-\frac{1}{2}t'Vt\right\} \times \left|\log \Psi(t) + \frac{1}{2}t'Vt - \frac{i^3}{6\sqrt{n}}\sum_j \sum_k \sum_m W_{jkm} t_j t_k t_m\right|$$

$$\leq \exp\left\{-\frac{1}{2}t'Vt\right\} \times \frac{1}{2}\left\{\frac{Wp^{3/2}}{6\sqrt{n}}\|t\|^3 + \frac{Rp^2}{2n}\|t\|^4 + \|t\|^2 O(n^{-1})\right\}^2$$

$$\times \exp\left[\left\{\left(\frac{Wp^{3/2}d_1}{6} + \frac{Rp^2 d_1^2}{2}\right) + O(n^{-1})\right\}\|t\|^2\right]$$

$$+ \exp\left\{-\frac{1}{2}t'Vt\right\} \times \left\{\frac{Rp^2}{2n}\|t\|^4 + \|t\|^2 O(n^{-1})\right\}$$

$$= \exp\left[-\frac{1}{2}t'\left\{V - \left(\frac{Wp^{3/2}d_1}{3} + Rp^2 d_1^2\right)I_n + O(n^{-1})\right\}t\right] \times \{\|t\|^6 O(n^{-1})\}$$

$$+ \exp\left\{-\frac{1}{2}t'Vt\right\} \times \{\|t\|^2 O(n^{-1}) + \|t\|^4 O(n^{-1})\}, \quad \text{for } \|t\| \leq d_1\sqrt{n}. \tag{3.1.13}$$

Since d_1 satisfies (3.1.8), then

$$v_0 - \frac{Wp^{3/2}d_1}{3} - Rp^2 d_1^2 > 0,$$

which implies the relation (3.1.9).

Lemma 3.1.3. *For every $\alpha > 0$,*

$$P(|Z_j| \geq n^\alpha) = O(n^{-\frac{1}{2}}), \tag{3.1.14}$$

for $j = 1, \ldots, p$.

Proof. Let $C^n_{j,k}$ be the kth order cumulant of Z_j. We may write

$$C^n_{j,k} = 2^{-k}n^{-\frac{k}{2}}\mathrm{cum}\left\{\mathbf{Y}'_n\Sigma_n^{\frac{1}{2}}G_{j-1}\Sigma_n^{\frac{1}{2}}\mathbf{Y}_n, \ldots, \mathbf{Y}'_n\Sigma_n^{\frac{1}{2}}G_{j-1}\Sigma_n^{\frac{1}{2}}\mathbf{Y}_n\right\},$$

where $\mathbf{Y}'_n = (Y_1, \ldots, Y_n)$ is a random vector distributed as $N(0, I_n)$. Denote the (l, m)-th component of $\Sigma_n^{1/2}G_{j-1}\Sigma_n^{1/2}$ by $C^{(j)}_{lm}$. Then using the fundamental properties of cumulant, we have

$$
\begin{aligned}
C^n_{j,k} &= 2^{-k}n^{-\frac{k}{2}}\sum_{l_1=1}^{n}\sum_{m_1=1}^{n}\cdots\sum_{l_k=1}^{n}\sum_{m_k=1}^{n}C^{(j)}_{l_1m_1}\cdots C^{(j)}_{l_km_k} \\
&\quad \times \mathrm{cum}\{Y_{l_1}Y_{m_1}, \ldots, Y_{l_k}Y_{m_k}\},
\end{aligned}
\tag{3.1.15}
$$

(see Theorem 2.3.1 of Brillinger(1975)). By Theorem 2.3.2 of Brillinger(1975), it is not difficult to show

$$C^n_{j,k} = O\left\{n^{-\frac{k}{2}}\mathrm{tr}(G_{j-1}\Sigma_n)^k\right\}.$$

By Lemma 3.1.1, we have $\mathrm{tr}(G_{j-1}\Sigma_n)^k = O(n)$, which implies

$$
C^n_{j,k} = \begin{cases} O(n^{-\frac{k}{2}+1}) & \text{for } k \geq 2 \\ 0 & \text{for } k = 1, \end{cases}
\tag{3.1.16}
$$

For general random variables $U_1, \ldots, U_k$ it is known that

$$E(U_1 \ldots U_k) = \sum_{\nu}\mathrm{cum}\{U_j(j \in \nu_1)\}\cdots\mathrm{cum}\{U_j(j \in \nu_p)\},
\tag{3.1.17}$$

where the summation is over all partitions $(\nu_1, \nu_2, \ldots, \nu_p)$ $(p = 1, \ldots, k)$ of integers $1, \ldots, k$ (see Brillinger(1975)).
From (3.1.16) and (3.1.17) it follows that

$$E(Z_j)^{2k} = O(1).
\tag{3.1.18}$$

By Tchebychev's inequality and choosing k so that $k \geq 1/4\alpha$, we have

$$P\left[|Z_j| \geq n^{\alpha}\right] \leq \frac{E(Z_j)^{2k}}{n^{2k\alpha}} = O(n^{-2k\alpha}) = O(n^{-\frac{1}{2}}),$$

which completes the proof.

We next mention a smoothing lemma. Let P_n and Φ_v be the probability distributions of $\mathbf{Z}$ and $N(0, V)$, respectively. We set down

$$L(f) = \sup\{|f(\mathbf{x}) - f(\mathbf{y})|\,;\, \mathbf{x}, \mathbf{y} \in R^p\},$$

for real-valued, Borel-measurable function f on R^p. Also we define

$$w_f(2\epsilon : \Phi_v) = \sup\left[\int \sup\{|f(\mathbf{x}_1 + \mathbf{y}) - f(\mathbf{x}_2 + \mathbf{y})|\,;\, \mathbf{x}_1, \mathbf{x}_2 \in B(\mathbf{x} : 2\epsilon)\}\,\Phi_v(d\mathbf{x})\,;\, \mathbf{y} \in R^p\right],$$

where $B(\mathbf{x} : 2\epsilon) = \{\mathbf{z}; \|\mathbf{z} - \mathbf{x}\| \le 2\epsilon\}$. Then we get the following smoothing lemma (see Bhattacharya and Rao(1976, p.97–98 and p.113)).

Lemma 3.1.4. *Let ϵ be a positive number. For every real-valued, bounded, Borel-measurable function f on R^p, there exists a kernel probability measure K_ϵ such that*

$$\left| \int f \, d(P_n - \Phi_v) \right| \le \frac{4}{3} \left[\frac{1}{2} L(f) \times \|(P_n - \Phi_v) \star K_\epsilon\| + w_f(2\epsilon : \Phi_v) \right],$$

where K_ϵ satisfies

$$K_\epsilon(B(\mathbf{0}, r)^c) = O\left\{ \left(\frac{\epsilon}{\gamma} \right)^3 \right\}, \tag{3.1.19}$$

and the Fourier transform $\hat{K}_\epsilon$ satisfies

$$\hat{K}_\epsilon(\mathbf{t}) = 0 \quad \text{for} \quad \|\mathbf{t}\| \ge \frac{8p^{4/3}}{\pi^{1/3}\epsilon}.$$

Here $\|(P_n - \Phi_v) \star K_\epsilon\|$ is the variation norm of the convolution of $(P_n - \Phi_v)$ and K_ϵ.

We may now state

Theorem 3.1.1. *Let P_n and Φ_v be the probability distributions of $\mathbf{Z}$ and $N(\mathbf{0}, V)$, respectively. Then for every real-valued, bounded, Borel-measurable function f on R^p,*

$$\left| \int f \, d(P_n - \Phi_v) \right| = O(n^{-\frac{1}{2}}) + \frac{4}{3} w_f \left(\frac{16p^{4/3}}{d_1 \pi^{1/3} \sqrt{n}} : \Phi_v \right),$$

where d_1 is defined in Lemma 3.1.2.

Proof. Putting $\epsilon = 8p^{4/3}/(d_1 \pi^{1/3} \sqrt{n})$ we can see that

$$\hat{K}_\epsilon(\mathbf{t}) = 0, \quad \text{for} \quad \|\mathbf{t}\| \ge d_1 \sqrt{n}, \tag{3.1.20}$$

For $B \in \mathbf{B}^p$, Borel set of R^p, define

$$\Psi_1(B) = \int_B N(\mathbf{y} : V) \left[\sum_j \sum_k \sum_m \frac{W_{jkm}}{6} H_{ijm}(\mathbf{y}) \right] d\mathbf{y},$$

where $\mathbf{y} = (y_1, \ldots, y_p)'$,

$$N(\mathbf{y} : V) = (2\pi)^{-\frac{p}{2}} |V|^{-\frac{1}{2}} \exp\left\{ -\frac{1}{2} \mathbf{y}' V^{-1} \mathbf{y} \right\},$$

and

$$H_{jkm}(\mathbf{y}) = \frac{(-1)}{N(\mathbf{y} : V)} \frac{\partial^3}{\partial y_j \partial y_k \partial y_m} N(\mathbf{y} : V),$$

and put $Q_n = P_n - \Phi_v - (1/\sqrt{n})\Psi_1$. Notice that

$$
\begin{aligned}
\|(P_n - \Phi_v) \star K_\epsilon\| &= 2\sup\{|(P_n - \Phi_v) \star K_\epsilon(B)| ; B \in \mathbf{B}^p\} \\
&\le 2\sup\{|(P_n - \Phi_v) \star K_\epsilon(B)| ; B \in \mathbf{B}^p \quad \text{and} \quad B \subset B(\mathbf{0}, \gamma_n)\} \\
&\quad + 2\sup\{|(P_n - \Phi_v) \star K_\epsilon(B)| ; B \in \mathbf{B}^p \quad \text{and} \quad B \subset B(\mathbf{0}, \gamma_n)^c\},
\end{aligned}
$$

where $\gamma_n = n^\alpha$, $0 \le \alpha \le 1/(2p)$. For $B \subset B(\mathbf{0}, \gamma_n)^c$, we have

$$
\begin{aligned}
|(P_n - \Phi_v) \star K_\epsilon(B)| &\le |P_n \star K_\epsilon(B)| + |\Phi_v \star K_\epsilon(B)| \\
&\le P\left(\|\mathbf{Z}\| \ge \frac{\gamma_n}{2}\right) + K_\epsilon\left(B\left(\mathbf{0}, \frac{\gamma_n}{2}\right)^c\right) + \int_{\|\mathbf{x}\| \ge \gamma_n/2} \Phi_v(d\mathbf{x}) + K_\epsilon\left(B\left(\mathbf{0}, \frac{\gamma_n}{2}\right)^c\right).
\end{aligned}
$$

It is easy to check

$$
\int_{\|\mathbf{x}\| \ge \gamma_n/2} \Phi_v(d\mathbf{x}) = O(n^{-\frac{1}{2}}).
$$

Lemma 3.1.3 implies

$$
P\left(\|\mathbf{Z}\| \ge \frac{\gamma_n}{2}\right) \le \sum_{j=1}^{p} P\left(|Z_j| \ge \frac{\gamma_n}{2p^{1/2}}\right) = O(n^{-\frac{1}{2}}).
$$

While (3.1.19) implies

$$
K_\epsilon\left(B\left(\mathbf{0}, \frac{\gamma_n}{2}\right)^c\right) = O(n^{-\frac{1}{2}}).
$$

Thus we have only to evaluate

$$
\sup\{|(P_n - \Phi_v) \star K_\epsilon(B)| \, ; B \subset B(\mathbf{0}, \gamma_n)\}.
$$

Here it holds that

$$
|(P_n - \Phi_v) \star K_\epsilon(B)| \le |Q_n \star K_\epsilon(B)| + n^{-\frac{1}{2}}|\Psi_1 \star K_\epsilon(B)|, \quad \text{for} \quad B \subset B(\mathbf{0}, \gamma_n).
$$

By Fourier inversion we have

$$
|Q_n \star K_\epsilon(B)| \le (2\pi)^{-p} \frac{\pi^{p/2} \gamma_n^p}{(p/2)\Gamma(p/2)} \int \left|\hat{Q}_n(\mathbf{t})\hat{K}_\epsilon(\mathbf{t})\right| \, d\mathbf{t}. \tag{3.1.21}
$$

By (3.1.20) and Lemma 3.1.2 the integral in the right hand side of (3.1.21) is dominated by

$$
\int_{\|\mathbf{t}\| \le d_1 \sqrt{n}} |\hat{Q}_n(\mathbf{t})| \, d\mathbf{t} = O(n^{-1}).
$$

Thus we get

$$
|Q_n \star K_\epsilon(B)| = O(n^{-\frac{1}{2}}), \quad \text{for} \quad B \subset B(\mathbf{0}, \gamma_n).
$$

Noting that

$$
|\Psi_1 \star K_\epsilon(B)| \le \frac{1}{2}\|\Psi_1\| \le \frac{1}{2}\int_{R^p} N(\mathbf{y} : V)\left|\sum_j \sum_k \sum_m \frac{W_{jkm}}{6} H_{jkm}(\mathbf{y})\right| d\mathbf{y},
$$

we have completed the proof.

Let Ξ be the class of all Borel-measurable convex subsets of R^p. If, $C \in \Xi$, we take $f = I_C$ (indicator function of C) in Theorem 3.1.1, we have

Theorem 3.1.2.

$$\sup_{C \in \Xi} |P_n(C) - \Phi_v(C)| = O(n^{-\frac{1}{2}}).$$

3.2. Validity of Edgeworth expansions of generalized maximum likelihood estimators for Gaussian ARMA processes

In this section we propose a generalized maximum likelihood estimator which includes the maximum likelihood estimator and the quasi-maximum likelihood estimator as special cases. Suppose that $\{X_t\}$ is a Gaussian ARMA process with spectral density $f_\theta(\lambda)$, where θ is an unknown parameter. Let $\hat{\theta}_n$ be the generalized maximum likelihood estimator of θ. Then we give the Edgeworth expansion of the distribution of $\hat{\theta}_n$ up to third order, and prove its validity. That is, as special cases we get the valid Edgeworth expansions for the maximum likelihood estimator and the quasi-maximum likelihood estimator which is defined by the value minimizing $\int_{-\pi}^{\pi} \{\log f_\theta(\lambda) + I_n(\lambda)/f_\theta(\lambda)\} \, d\lambda$ with respect to θ, where $I_n(\lambda)$ is the periodogram. We also consider the transformed statistic $\hat{\theta}_T = \hat{\theta}_n + (1/n)T(\hat{\theta}_n)$, where $T(\cdot)$ is a smooth function. Then we give the valid Edgeworth expansion for $\hat{\theta}_T$. By this Edgeworth expansion we can see that our generalized maximum likelihood estimator is always second-order asymptotically efficient in the class of second-order asymptotically median unbiased estimators. Moreover, the third-order asymptotic comparisons among generalized maximum likelihood estimators will be given.

Let $\mathbf{D}_d$ and $\mathbf{D}_{ARMA}^c$ be spaces of functions on $[-\pi, \pi]$ defined by

$$\mathbf{D}_d = \left\{ f : f(\lambda) = \sum_{u=-\infty}^{\infty} a(u) \exp(-iu\lambda), \ a(u) = a(-u), \right.$$

$$\left. \sum_{u=-\infty}^{\infty} (1+|u|)|a(u)| < d, \ \text{for some} \ d < \infty \right\},$$

$$\mathbf{D}_{ARMA}^c = \left\{ f : f(\lambda) = \frac{\sigma^2}{2\pi} \frac{\left|\sum_{j=0}^{q} a_j e^{ij\lambda}\right|^2}{\left|\sum_{j=0}^{p} b_j e^{ij\lambda}\right|^2}, \ (\sigma^2 > 0), \right.$$

$$\left. \underline{c} \le \frac{\sigma^2}{2\pi} \frac{\left|\sum_{j=0}^{q} a_j z^j\right|^2}{\left|\sum_{j=0}^{p} b_j z^j\right|^2} \le \overline{c}, \ \text{for} \ |z| \le 1, \ 0 < \underline{c} < \overline{c} < \infty \right\}.$$

We set down the following assumptions.

Assumption 3.2.1. The process $\{X_t; t = 0, \pm 1, \ldots\}$ is a Gaussian stationary process with the spectral density $f_{\theta_0}(\lambda) \in \mathbf{D}_{ARMA}^c$, $\theta_0 \in C \subset \Theta \subset R^1$, and mean 0. Here Θ is an open set of R^1 and C is a compact subset of Θ.

Assumption 3.2.2. The spectral density $f_\theta(\lambda)$ is continuously five times differentiable with respect to $\theta \in \Theta$, and the derivatives $\partial f_\theta/\partial\theta$, $\partial^2 f_\theta/\partial\theta^2$, $\partial^3 f_\theta/\partial\theta^3$, $\partial^4 f_\theta/\partial\theta^4$ and $\partial^5 f_\theta/\partial\theta^5$ belong to $\mathbf{D}_d$.

Assumption 3.2.3. There exists $d_0 > 0$ such that

$$I(\theta) = \frac{1}{4\pi} \int_{-\pi}^{\pi} \left\{ \frac{\partial}{\partial \theta} \log f_\theta(\lambda) \right\}^2 d\lambda \geq d_0 > 0, \quad \text{for all} \quad \theta \in \Theta.$$

Suppose that a stretch $\mathbf{X}_n = (X_1, \ldots, X_n)'$ of the series $\{X_t\}$ is available. Let $\Sigma_n = \Sigma_n(\theta_0)$ be the covariance matrix of $\mathbf{X}_n$. The (m, l)-th element of Σ_n is given by $\int_{-\pi}^{\pi} \exp\{i(m-l)\lambda\} f_{\theta_0}(\lambda) \, d\lambda$. Let $A_n(\theta)$ and $B_n(\theta)$ be $n \times n$ Toeplitz matrices associated with harmonic functions $g_\theta(\lambda)$ and $h_\theta(\lambda)$, where $g_\theta \in \mathbf{D}_{ARMA}^c$, $h_\theta \in \mathbf{D}_d$ (i.e., the (m, l)-th elements of $A_n(\theta)$ and $B_n(\theta)$ are given by $\int_{-\pi}^{\pi} \exp\{i(m-l)\lambda\} g_\theta(\lambda) \, d\lambda$ and $\int_{-\pi}^{\pi} \exp\{i(m-l)\lambda\} h_\theta(\lambda) \, d\lambda$, respectively). We impose further assumptions.

Assumption 3.2.4. The functions g_θ and h_θ are continuously four times differentiable with respect to $\theta \in \Theta$, and the derivatives $\partial g_\theta/\partial\theta, \ldots, \partial^4 g_\theta/\partial\theta^4$, $\partial h_\theta/\partial\theta, \ldots, \partial^4 h_\theta/\partial\theta^4$ $\theta \in \Theta$, belong to $\mathbf{D}_d$. Also g_θ and h_θ satisfy

$$g_\theta(\lambda)^{-2} h_\theta(\lambda) = \frac{1}{2} f_\theta(\lambda)^{-2} \frac{\partial}{\partial\theta} f_\theta(\lambda). \tag{3.2.1}$$

Assumption 3.2.5. A function $b_n(\theta)$ is four times continuously differentiable with respect to θ, and is written as

$$b_n(\theta) = \frac{1}{4\pi} \int_{-\pi}^{\pi} f_\theta(\lambda)^{-1} \frac{\partial}{\partial\theta} f_\theta(\lambda) \, d\lambda + O(n^{-1}).$$

Now consider the following equation;

$$\frac{1}{n} \mathbf{X}_n' A_n(\theta)^{-1} B_n(\theta) A_n(\theta)^{-1} \mathbf{X}_n = b_n(\theta), \quad \theta \in \Theta \tag{3.2.2}$$

A generalized maximum likelihood estimator $\hat{\theta}_n$ of θ_0 is defined by a value of θ that satisfies the equation (3.2.2). This estimator $\hat{\theta}_n$ includes the following cases;

Example 3.2.1. Put $g_\theta = f_\theta$, $h_\theta = \frac{1}{2}\partial f_\theta/\partial\theta$ and $b_n(\theta) = \frac{1}{2n}\mathrm{tr}\Sigma_n^{-1}\frac{\partial}{\partial\theta}\Sigma_n$, then by Theorem 2.2.1,

$$b_n(\theta) = \frac{1}{4\pi} \int_{-\pi}^{\pi} f_\theta^{-1} \frac{\partial}{\partial\theta} f_\theta \, d\lambda + O(n^{-1}).$$

The estimator $\hat{\theta}_n$ becomes the maximum likelihood estimator (see (2.2.7) and (2.2.29)).

Example 3.2.2. Put $g_\theta = \frac{1}{2\pi}$, $h_\theta = \frac{1}{8\pi^2}\frac{\partial}{\partial\theta}f_\theta \cdot f_\theta^{-2}$ and $b_n(\theta) = \frac{1}{4\pi}\int_{-\pi}^{\pi}(\frac{\partial}{\partial\theta}f_\theta)f_\theta^{-1}\,d\lambda$. Then (3.2.2) is written as

$$\mathbf{X}_n' \begin{pmatrix} & & l & \\ & & \vdots & \\ m & \cdots & \int_{-\pi}^{\pi} e^{i(m-l)\lambda}\frac{1}{8\pi^2 n}\left(\frac{\partial}{\partial\theta}f_\theta\right)f_\theta^{-2}\,d\lambda & \end{pmatrix} \mathbf{X}_n = \frac{1}{4\pi}\int_{-\pi}^{\pi} f_\theta^{-1}\frac{\partial}{\partial\theta}f_\theta\,d\lambda. \tag{3.2.3}$$

We can see that the equation (3.2.3) is equivalent to

$$\frac{\partial}{\partial \theta} \int_{-\pi}^{\pi} \left\{ \log f_\theta(\lambda) + \frac{I_n(\lambda)}{f_\theta(\lambda)} \right\} d\lambda = 0,$$

where $I_n(\lambda) = (1/2\pi n)|\sum_{t=1}^{n} X_t e^{it\lambda}|^2$. Thus the estimator $\hat{\theta}_n$ becomes the quasi-maximum likelihood estimator (see Dunsmuir and Hannan(1976) and Hosoya and Taniguchi(1982)).

Initially, we present the following basic theorem which is useful for the higher order asymptotic theory in time series analysis.

Theorem 3.2.1. *Assume that Assumptions 3.2.1 – 3.2.5 hold. Let α be an arbitrary fixed number such that $0 < \alpha < \frac{3}{8}$.*
(1) *There exists a statistic $\hat{\theta}_n$ which solves (3.2.2) such that for some $d_1 > 0$,*

$$P_{\theta_0}^n \left[|\hat{\theta}_n - \theta_0| < d_1 n^{\alpha - 1/2} \right] = 1 - o(n^{-1}), \tag{3.2.4}$$

uniformly for $\theta_0 \in C$.
(2) *For $\{\hat{\theta}_n\}$ satisfying (3.2.4),*

$$\sup_{B \in \mathbf{B}_0} \left| P_{\theta_0}^n \left[\{nI(\theta_0)\}^{1/2} (\hat{\theta}_n - \theta_0) \in B \right] - \int_B \phi(x) p_3^n(x) \, dx \right| = o(n^{-1}), \tag{3.2.5}$$

uniformly for $\theta_0 \in C$, where $\mathbf{B}_0$ is a class of Borel sets of R^1 satisfying

$$\sup_{B \in \mathbf{B}_0} \int_{(\partial B)^\epsilon} \phi(x) p_3^n(x) \, dx = O(\epsilon). \tag{3.2.6}$$

Here $\phi(x) = \frac{1}{\sqrt{2\pi}} \exp\{-\frac{1}{2} x^2\}$, and $p_3^n(x) = 1 + \frac{1}{\sqrt{n}} q(x) + \frac{1}{n} \gamma(x)$ where $q(x)$ and $\gamma(x)$ are polynomials.

Later we shall give the coefficients of $q(x)$ and $\gamma(x)$ explicitly by using the spectral density f_θ. Before proving Theorem 3.2.1, we state some preparation and several lemmas. We set down

$$l_n(\theta) = \mathbf{X}_n' G_n(\theta) \mathbf{X}_n - n b_n(\theta),$$

where $G_n(\theta) = A_n(\theta)^{-1} B_n(\theta) A_n(\theta)^{-1}$. Let

$$Z_1(\theta) = \frac{1}{\sqrt{n}} \{ \mathbf{X}_n' G_n(\theta) \mathbf{X}_n - n b_n(\theta) \}, \tag{3.2.7}$$

$$Z_2(\theta) = \frac{1}{\sqrt{n}} \{ \mathbf{X}_n' \dot{G}_n(\theta) \mathbf{X}_n - \mathrm{tr} \Sigma_n(\theta) \dot{G}_n(\theta) \}, \tag{3.2.8}$$

$$Z_3(\theta) = \frac{1}{\sqrt{n}} \{ \mathbf{X}_n' \ddot{G}_n(\theta) \mathbf{X}_n - \mathrm{tr} \Sigma_n(\theta) \ddot{G}_n(\theta) \}, \tag{3.2.9}$$

where $\dot{G}_n(\theta) = \frac{\partial}{\partial \theta} G_n(\theta)$ and $\ddot{G}_n(\theta) = \frac{\partial^2}{\partial \theta^2} G_n(\theta)$. Henceforth, for simplicity, we sometimes use A, B, G, Σ, Z_1, Z_2 and Z_3 instead of $A_n(\theta)$, $B_n(\theta)$, $G_n(\theta)$, $\Sigma_n(\theta)$, $Z_1(\theta)$, $Z_2(\theta)$ and $Z_3(\theta)$, respectively. It is easy to show that

$$\dot{G} = -A^{-1}\dot{A}A^{-1}BA^{-1} - A^{-1}BA^{-1}\dot{A}A^{-1} + A^{-1}\dot{B}A^{-1}, \tag{3.2.10}$$

$$\begin{aligned}
\ddot{G} = \ & A^{-1}\dot{A}A^{-1}\dot{A}A^{-1}BA^{-1} + A^{-1}BA^{-1}\dot{A}A^{-1}\dot{A}A^{-1} \\
& - A^{-1}\ddot{A}A^{-1}BA^{-1} - A^{-1}BA^{-1}\ddot{A}A^{-1} + A^{-1}\dot{A}A^{-1}\dot{A}A^{-1}BA^{-1} + A^{-1}BA^{-1}\dot{A}A^{-1}\dot{A}A^{-1} \\
& - A^{-1}\dot{A}A^{-1}\dot{B}A^{-1} - A^{-1}\dot{B}A^{-1}\dot{A}A^{-1} + 2A^{-1}\dot{A}A^{-1}BA^{-1}\dot{A}A^{-1} - A^{-1}\dot{A}A^{-1}\dot{B}A^{-1} \\
& - A^{-1}\dot{B}A^{-1}\dot{A}A^{-1} + A^{-1}\ddot{B}A^{-1},
\end{aligned} \tag{3.2.11}$$

where $\dot{A} = \frac{\partial}{\partial\theta}A_n(\theta)$, $\ddot{A} = \frac{\partial^2}{\partial\theta^2}A_n(\theta)$, $\dot{B} = \frac{\partial}{\partial\theta}B_n(\theta)$ and $\ddot{B} = \frac{\partial^2}{\partial\theta^2}B_n(\theta)$.

Since the generalized maximum likelihood estimator is approximated by simple functions of Z_1, Z_2 and Z_3. To give the asymptotic expansion we must evaluate the asymptotic cumulants (moments) of Z_1, Z_2 and Z_3. Remembering Theorem 2.2.1 we can show

Lemma 3.2.1. *Under Assumptions 3.2.1 – 3.2.5,*

$$E_\theta\{Z_1(\theta)\}^2 = I(\theta) + O(n^{-1}),$$

$$E_\theta\{Z_1(\theta)Z_2(\theta)\} = J(\theta) + O(n^{-1}),$$

$$E_\theta\{Z_1(\theta)\}^3 = \frac{1}{\sqrt{n}}K(\theta) + \frac{3}{\sqrt{n}}I(\theta)\mu(\theta) + O(n^{-3/2}),$$

$$E_\theta\{Z_1(\theta)Z_3(\theta)\} = L(\theta) + O(n^{-1}),$$

$$E_\theta\{Z_2(\theta)\}^2 = M(\theta) + O(n^{-1}),$$

$$E_\theta\{Z_1(\theta)^2 Z_2(\theta)\} = \frac{1}{\sqrt{n}}N(\theta) + \frac{2}{\sqrt{n}}J(\theta)\mu(\theta) + O(n^{-3/2}),$$

$$\mathrm{cum}_\theta\{Z_1(\theta), Z_1(\theta), Z_1(\theta), Z_1(\theta)\} = \frac{1}{n}H(\theta) + O(n^{-2}),$$

$$E_\theta\left\{\frac{1}{n}\frac{\partial}{\partial\theta}l_n(\theta)\right\} = -I(\theta) + O(n^{-1}), \tag{3.2.12}$$

$$E_\theta\left\{\frac{1}{n}\frac{\partial^2}{\partial\theta^2}l_n(\theta)\right\} = -3J(\theta) - K(\theta) + O(n^{-1}), \tag{3.2.13}$$

$$E_\theta\left\{\frac{1}{n}\frac{\partial^3}{\partial\theta^3}l_n(\theta)\right\} = -4L(\theta) - 3M(\theta) - 6N(\theta) - H(\theta) + O(n^{-1}), \tag{3.2.14}$$

where $J(\theta)$, $K(\theta)$, $L(\theta)$, $M(\theta)$, $N(\theta)$ and $H(\theta)$ are given in Lemma 2.3.1 (for vector case) in terms of $f_\theta(\lambda)$, and $\mu(\theta)$ is defined by the relation

$$E_\theta\{Z_1(\theta)\} = \frac{1}{\sqrt{n}}\mu(\theta) + o(n^{-1}). \tag{3.2.15}$$

Here it may be noted that the asymptotic moments of the fundamental quantities Z_1, Z_2 and Z_3 coincide with those based on the exact likelihood except for the bias term $\mu(\theta)$ (see Lemma 2.3.1).

Put $\Omega_n = \Lambda_1^{-1}\Gamma_1\Lambda_2^{-1}\cdots\Gamma_{s-1}\Lambda_s^{-1}$, where $\Gamma_1,\ldots,\Gamma_{s-1}$, $\Lambda_1,\ldots,\Lambda_s$ are $n \times n$ Toeplitz type matrices associated with some harmonic functions $u_\theta^{(1)}(\lambda) \in \mathbf{D}_d,\ldots,u_\theta^{(s-1)}(\lambda) \in \mathbf{D}_d$, $v_\theta^{(1)}(\lambda) \in \mathbf{D}_{ARMA}^c,\ldots,v_\theta^{(s)}(\lambda) \in \mathbf{D}_{ARMA}^c$, respectively. Then we have

Lemma 3.2.2. *Under Assumptions 3.2.1, for every $\beta > 0$, and some $d_2 > 0$, we have*

$$P_\theta^n\left[\frac{1}{\sqrt{n}}|\mathbf{X}_n'\Omega_n\mathbf{X}_n - E_\theta(\mathbf{X}_n'\Omega_n\mathbf{X}_n)| > d_2 n^\beta\right] = o(n^{-1}) \tag{3.2.16}$$

uniformly for $\theta \in \Theta$.

Proof. Choose an integer $\eta \geq 1$ so that $2\eta\beta > 1$. By Tchebychev's inequality we have

$$P_\theta^n\left[\frac{1}{\sqrt{n}}|\mathbf{X}_n'\Omega_n\mathbf{X}_n - E_\theta(\mathbf{X}_n'\Omega_n\mathbf{X}_n)| > d_2 n^\beta\right]$$

$$\leq \ E_\theta \left[\frac{1}{\sqrt{n}} |\mathbf{X}'_n \Omega_n \mathbf{X}_n - E_\theta(\mathbf{X}'_n \Omega_n \mathbf{X}_n)| \right]^{2\eta} \Bigg/ (d_2 n^\beta)^{2\eta}. \qquad (3.2.17)$$

As in the proof of Lemma 3.1.3, we can show

$$E_\theta \left[\frac{1}{\sqrt{n}} |\mathbf{X}'_n \Omega_n \mathbf{X}_n - E_\theta(\mathbf{X}'_n \Omega_n \mathbf{X}_n)| \right]^{2\eta} = O(1), \qquad (3.2.18)$$

which implies (3.2.16) together with (3.2.17).

The following lemma is essentially due to Chibisov(1972).

Lemma 3.2.3. *Let Y_n be a random variable which has the stochastic expansion*

$$Y_n = Y_n^{(3)} + n^{-\frac{3}{2}} \xi_n, \qquad (3.2.19)$$

where the distribution of $Y_n^{(3)}$ has the Edgeworth expansion:

$$P\left[Y_n^{(3)} \in B \right] = \int_B \phi(x) p_3^n(x) \, dx + o(n^{-1}),$$

where B is a Borel set of R^1 satisfying (3.2.6).
Also ξ_n satisfies

$$P\left[|\xi_n| > \rho_n \sqrt{n} \right] = o(n^{-1}), \qquad (3.2.20)$$

where $\rho_n \to 0$, $\rho_n \sqrt{n} \to \infty$ as $n \to \infty$. Then

$$P[Y_n \in B] = \int_B \phi(x) p_3^n(x) \, dx + o(n^{-1}), \qquad (3.2.21)$$

for $B \in \mathbf{B}_0$.

We return to the proof of (1) in Theorem 3.2.1.

Proof of (1) in Theorem 3.2.1. Here we develop the discussion by using the argument similar to that of Bhattacharya and Ghosh(1978) and Taniguchi(1986a). Consider the equation

$$0 = n^{-1} l_n(\theta_0) + n^{-1}(\theta - \theta_0) \frac{\partial}{\partial \theta} l_n(\theta_0) + (2n)^{-1}(\theta - \theta_0)^2 \frac{\partial^2}{\partial \theta^2} l_n(\theta_0)$$

$$+ (6n)^{-1}(\theta - \theta_0)^3 \frac{\partial^3}{\partial \theta^3} l_n(\theta_0) + R_n(\theta), \qquad (3.2.22)$$

where $R_n(\theta)$ is the usual remainder in the Taylor expansion, for which it holds that

$$|R_n(\theta)| \leq \frac{1}{24n} |\theta - \theta_0|^4 \sup_{|\theta' - \theta| \leq |\theta - \theta_0|} \left| \frac{\partial^4}{\partial \theta^4} l_n(\theta') \right|. \qquad (3.2.23)$$

In view of Lemma 3.2.2, we can see that for every $\alpha > 0$ there exist positive constants d_3 and d_4 such that

$$P_{\theta_0}^n \left[|Z_1(\theta_0)| > d_3 n^\alpha \right] = o(n^{-1}), \qquad (3.2.24)$$

$$P_{\theta_0}^n \left[|Z_2(\theta_0)| > d_3 n^\alpha \right] = o(n^{-1}), \qquad (3.2.25)$$

$$P_{\theta_0}^n \left[|Z_3(\theta_0)| > d_3 n^\alpha \right] = o(n^{-1}), \qquad (3.2.26)$$

$$P_{\theta_0}^n \left[\left| \frac{1}{\sqrt{n}} \left\{ \frac{\partial^3}{\partial \theta^3} l_n(\theta_0) - E_{\theta_0} \frac{\partial^3}{\partial \theta^3} l_n(\theta_0) \right\} \right| > d_3 n^\alpha \right] = o(n^{-1}), \qquad (3.2.27)$$

$$P_{\theta_0}^n \left[|R_n(\theta)| > |\theta - \theta_0|^4 \{ d_4 n^\alpha \} \right] = o(n^{-1}). \qquad (3.2.28)$$

Therefore, on a set having $P_{\theta_0}^n$-probability at least $1 - o(n^{-1})$, for some constants $d_5 > 0$ and $d_6 > 0$ we can rewrite (3.2.22) as

$$
\begin{aligned}
\theta - \theta_0 = \{I(\theta_0) + \eta_n\}^{-1} & \left[\delta_n + (2n)^{-1}(\theta - \theta_0)^2 \frac{\partial^2}{\partial \theta^2} l_n(\theta_0) \right. \\
& \left. + (6n)^{-1}(\theta - \theta_0)^3 \frac{\partial^3}{\partial \theta^3} l_n(\theta_0) + d_5 |\theta - \theta_0|^4 \zeta_n \right]
\end{aligned}
\tag{3.2.29}
$$

where η_n and δ_n are random variables whose absolute values are less than $d_6 n^{-1/2+\alpha}$ and ζ_n is a random variable whose absolute value is less than $d_4 n^\alpha$. There exist a sufficiently large $d_7 > 0$ and an integer n_0 such that if $n > n_0$ and $|\theta - \theta_0| \leq d_7 n^{-1/2+\alpha}$ $(0 < \alpha < 3/8)$, the right-hand side of (3.2.29) is less than $d_7 n^{-1/2+\alpha}$. Applying the Brouwer fixed point theorem to the right-hand side of (3.2.29) we have proved (1) of Theorem 3.2.1.

Now we set down

$$
V_n = \sqrt{n}(\hat{\theta}_n - \theta_0),
$$

$$
I_n(\theta) = -\frac{1}{n} E_\theta \left\{ \frac{\partial}{\partial \theta} l_n(\theta) \right\},
$$

and

$$
\begin{aligned}
U_n(\theta) = & \frac{Z_1(\theta)}{I_n(\theta)} + \frac{Z_1(\theta)Z_2(\theta)}{I(\theta)^2 \sqrt{n}} - \frac{3J(\theta) + K(\theta)}{2I(\theta)^3 \sqrt{n}} Z_1(\theta)^2 \\
& + \frac{1}{I(\theta)^3 n} \left[Z_1(\theta)Z_2(\theta)^2 + \frac{1}{2} Z_1(\theta)^2 Z_3(\theta) \right. \\
& - \frac{3\{3J(\theta) + K(\theta)\}}{2I(\theta)} Z_1(\theta)^2 Z_2(\theta) + \frac{\{3J(\theta) + K(\theta)\}^2}{2I(\theta)^2} Z_1(\theta)^3 \\
& \left. - \frac{4L(\theta) + 3M(\theta) + 6N(\theta) + H(\theta)}{6I(\theta)} Z_1(\theta)^3 \right].
\end{aligned}
$$

Lemma 3.2.4. *Under Assumptions 3.2.1 – 3.2.5, we have the following stochastic expansion*

$$
\sqrt{n}(\hat{\theta}_n - \theta_0) = U_n(\theta_0) + n^{-\frac{3}{2}} \zeta_n,
\tag{3.2.30}
$$

where ζ_n satisfies $P_{\theta_0}^n [|\zeta_n| > \rho_n \sqrt{n}] = o(n^{-1})$ for some sequence $\rho_n \to 0$, $\rho_n \sqrt{n} \to \infty$ as $n \to \infty$.

Proof. From the equation $l_n(\hat{\theta}_n) = 0$, we have

$$
\begin{aligned}
0 = & Z_1(\theta_0) + n^{-\frac{1}{2}} Z_2(\theta_0) V_n - I_n(\theta_0) V_n \\
& + \frac{1}{2} n^{-\frac{3}{2}} \left\{ \frac{\partial^2}{\partial \theta^2} l_n(\theta_0) \right\} V_n^2 + \frac{1}{6n^2} \left\{ \frac{\partial^3}{\partial \theta^3} l_n(\theta_0) \right\} V_n^3 \\
& + \frac{1}{24 n^{5/2}} \left\{ \frac{\partial^4}{\partial \theta^4} l_n(\theta_\star) \right\} V_n^4,
\end{aligned}
\tag{3.2.31}
$$

where $|\theta \star -\theta_0| \leq |\hat{\theta}_n - \theta_0|$. We rewrite (3.2.31) as

$$
\begin{aligned}
V_n =\ & \frac{Z_1(\theta_0)}{I_n(\theta_0)} + \frac{1}{I_n(\theta_0)\sqrt{n}} Z_2(\theta_0) V_n + \frac{1}{2I_n(\theta_0)\sqrt{n}} \left\{ \frac{1}{n} \frac{\partial^2}{\partial \theta^2} l_n(\theta_0) \right\} V_n^2 \\
& + \frac{1}{6I_n(\theta_0)n} \left\{ \frac{1}{n} \frac{\partial^3}{\partial \theta^3} l_n(\theta_0) \right\} V_n^3 + \frac{1}{24I_n(\theta_0)n\sqrt{n}} \left\{ \frac{1}{n} \frac{\partial^4}{\partial \theta^4} l_n(\theta\star) \right\} V_n^4.
\end{aligned} \tag{3.2.32}
$$

Noting (3.2.4), (3.2.24) – (3.2.28) with $0 < \alpha < 1/10$, we can write (3.2.32) as

$$
V_n = \frac{Z_1}{I_n} + \frac{1}{\sqrt{n}} a_n(1), \tag{3.2.33}
$$

where $P_{\theta_0}^n[|a_n(1)| > d_8 n^{2\alpha}] = o(n^{-1})$, for some $d_8 > 0$. Substituting (3.2.33) for the right-hand side of (3.2.32), and noting (3.2.13) we have

$$
V_n = \frac{Z_1}{I_n} + \frac{Z_1 Z_2}{I^2 \sqrt{n}} - \frac{3J + K}{2I^3 \sqrt{n}} Z_1^2 + \frac{1}{n} a_n(2), \tag{3.2.34}
$$

where $P_{\theta_0}^n[|a_n(2)| > d_9 n^{3\alpha}] = o(n^{-1})$, for some $d_9 > 0$. Again substituting (3.2.34) for the right-hand side of (3.2.32), and noting (3.2.14) we have

$$
V_n = U_n(\theta_0) + n^{-\frac{3}{2}} \zeta_n, \tag{3.2.35}
$$

where $P_{\theta_0}^n[|\zeta_n| > d_{10} n^{5\alpha}] = o(n^{-1})$, for some $d_{10} > 0$. Since $0 < \alpha < 1/10$, we have the desired result.

Remark 3.2.1. By Lemma 3.2.3, the Edgeworth expansion for $\sqrt{n}(\hat{\theta}_n - \theta_0)$ (up to order n^{-1}) is equal to that for $U_n(\theta_0)$ on $B \in \mathbf{B}_0$. Thus we have only to derive the Edgeworth expansion for $U_n(\theta_0)$.

To derive the Edgeworth expansion for $U_n(\theta_0)$ we need that for $\mathbf{Z} = (Z_1(\theta), Z_2(\theta), Z_3(\theta))'$. For this we give an asymptotic expansion of the characteristic function of $\mathbf{Z}$. Put

$$
\tau(\mathbf{t}) = E_\theta \{ e^{i\mathbf{t}'\mathbf{Z}} \},
$$

where $\mathbf{t} = (t_1, t_2, t_3)'$. Then it is easy to show

$$
\begin{aligned}
\tau(\mathbf{t}) =\ & \det \left\{ I(n \times n) - \frac{2i}{\sqrt{n}} \Sigma^{\frac{1}{2}} (t_1 G + t_2 \dot{G} + t_3 \ddot{G}) \Sigma^{\frac{1}{2}} \right\}^{-\frac{1}{2}} \\
& \times \exp -\frac{i}{\sqrt{n}} (t_1 n b_n(\theta) + t_2 \text{tr} \dot{G} \Sigma + t_3 \text{tr} \ddot{G} \Sigma),
\end{aligned} \tag{3.2.36}
$$

where $I(n \times n)$ is the $n \times n$ identity matrix. Let ρ_j be the jth latent root of $S = \Sigma^{1/2}(t_1 G + t_2 \dot{G} + t_3 \ddot{G})\Sigma^{1/2}$ ($\rho_1^2 \geq \ldots \geq \rho_n^2 \geq 0$). Of course each ρ_j is a real number. Then we have

$$
\log \tau(\mathbf{t}) = -\frac{1}{2} \sum_{j=1}^n \log \left\{ 1 - \frac{2i}{\sqrt{n}} \rho_j \right\} - \frac{i}{\sqrt{n}} (t_1 n b_n(\theta) + t_2 \text{tr} \dot{G} \Sigma + t_3 \text{tr} \ddot{G} \Sigma). \tag{3.2.37}
$$

Note the relation

$$
\log(1 - ih) = -ih + \frac{h^2}{2} + \frac{ih^3}{3} - \frac{h^4}{4} - \frac{ih^5}{5} + h^6 \int_0^1 (1-v)^5 \frac{dv}{(1-ivh)^6}, \tag{3.2.38}
$$

where

$$\left| \int_0^1 (1-v)^5 \frac{dv}{(1-ivh)^6} \right| \leq 1$$

(e.g., Bhattacharya and Rao(1976, p.57)). By (3.2.38), the relation (3.2.37) is

$$\log \tau(\mathbf{t}) = -\frac{1}{2} \sum_{j=1}^n \left[-\frac{2i\rho_j}{\sqrt{n}} + \frac{4\rho_j^2}{2n} + \frac{8i\rho_j^3}{3n\sqrt{n}} - \frac{16\rho_j^4}{4n^2} - \frac{2^5 i \rho_j^5}{5n^{5/2}} + \frac{2^6 \rho_j^6 \gamma_j}{n^3} \right]$$
$$-\frac{i}{\sqrt{n}}(t_1 n b_n(\theta) + t_2 \mathrm{tr} \dot{G} \Sigma + t_3 \mathrm{tr} \ddot{G} \Sigma), \tag{3.2.39}$$

where $|\gamma_j| \leq 1$. Remembering (3.2.15) we have

$$\log \tau(\mathbf{t}) = it_1 \left\{ \frac{\mu(\theta)}{\sqrt{n}} + o(n^{-1}) \right\} + \frac{i^2}{n} \mathrm{tr} S^2 + \frac{4i^2}{3n\sqrt{n}} \mathrm{tr} S^3$$
$$+\frac{2i^4}{n^2} \mathrm{tr} S^4 + \frac{16i^5}{5n^{5/2}} \mathrm{tr} S^5 + R_6, \tag{3.2.40}$$

where $|R_6| \leq \frac{2}{n^3} \mathrm{tr} S^6$. Using Theorem 2.2.1 we have the following expressions;

$$\frac{2i^2}{n} \mathrm{tr} S^2 = \sum_{j=1}^3 \sum_{k=1}^3 \left\{ A_{jk} + \frac{B_{jk}}{n} + O(n^{-\frac{3}{2}}) \right\} (it_j)(it_k), \tag{3.2.41}$$

$$\frac{8i^3}{n} \mathrm{tr} S^3 = \sum_{j=1}^3 \sum_{k=1}^3 \sum_{l=1}^3 \{A_{jkl} + O(n^{-1})\}(it_j)(it_k)(it_l), \tag{3.2.42}$$

$$\frac{48i^4}{n} \mathrm{tr} S^4 = \sum_{j=1}^3 \sum_{k=1}^3 \sum_{l=1}^3 \sum_{m=1}^3 \{A_{jklm} + O(n^{-1})\}(it_j)(it_k)(it_l)(it_m), \tag{3.2.43}$$

$$\frac{384i^5}{n} \mathrm{tr} S^5 = \sum_{j=1}^3 \sum_{k=1}^3 \sum_{l=1}^3 \sum_{m=1}^3 \sum_{r=1}^3 \{A_{jklmr} + O(n^{-1})\}$$
$$\times (it_j)(it_k)(it_l)(it_m)(it_r), \tag{3.2.44}$$

$$\frac{i^6}{n} \mathrm{tr} S^6 = \sum_{j=1}^3 \sum_{k=1}^3 \sum_{l=1}^3 \sum_{m=1}^3 \sum_{r=1}^3 \sum_{s=1}^3 \{A_{jklmrs} + O(n^{-1})\}$$
$$\times (it_j)(it_k)(it_l)(it_m)(it_r)(it_s). \tag{3.2.45}$$

For examples we can see that $A_{11} = I(\theta)$, $A_{12} = J(\theta)$, $A_{13} = L(\theta)$, $A_{22} = M(\theta)$, $A_{111} = K(\theta)$, $A_{112} = N(\theta)$, $A_{1111} = H(\theta)$, etc. Thus (3.2.40) is written as

$$\log \tau(\mathbf{t}) = it_1 \left\{ \frac{1}{\sqrt{n}} \mu(\theta) + o(n^{-1}) \right\}$$
$$+\frac{1}{2} \sum_{j=1}^3 \sum_{k=1}^3 \left\{ A_{jk} + \frac{B_{jk}}{n} + O(n^{-\frac{3}{2}}) \right\} (it_j)(it_k)$$
$$+\frac{1}{6\sqrt{n}} \sum_{j=1}^3 \sum_{k=1}^3 \sum_{l=1}^3 \{A_{jkl} + O(n^{-1})\}(it_j)(it_k)(it_l)$$
$$+\frac{1}{24n} \sum_{j=1}^3 \sum_{k=1}^3 \sum_{l=1}^3 \sum_{m=1}^3 \{A_{jklm} + O(n^{-1})\}(it_j)(it_k)(it_l)(it_m)$$
$$+\frac{1}{120n^{3/2}} \sum_{j=1}^3 \sum_{k=1}^3 \sum_{l=1}^3 \sum_{m=1}^3 \sum_{r=1}^3 \{A_{jklmr} + O(n^{-1})\}(it_j)(it_k)(it_l)(it_m)(it_r)$$
$$+R_6. \tag{3.2.46}$$

We set down $\Omega = \{A_{jk}\}$, 3×3 matrix, and $\|\mathbf{t}\| = \sqrt{t_1^2 + t_2^2 + t_3^2}$. If Ω is singular it is not difficult to show that

$$
\begin{aligned}
Z_1(\theta) &= c_1(\theta)Z_2(\theta) + d_1(\theta) \\
&= c_2(\theta)Z_3(\theta) + d_2(\theta),
\end{aligned}
$$
$$
\text{in} \quad P, \quad \text{for some constants} \quad c_i(\theta), d_i(\theta) \quad (i = 1, 2) \tag{3.2.47}
$$

which implies that the limiting distribution of $\mathbf{Z}$ is reduced to that of Z_1. Thus, without loss of generality, henceforth we consider the case when Ω is nonsingular.

Lemma 3.2.5. *If we take n sufficiently large, then for a $\delta_1 > 0$ and for all $\mathbf{t}$ satisfying $\|\mathbf{t}\| \le \delta_1\sqrt{n}$, there exists a positive definite matrix Q_0 and polynomial functions $F_1(\cdot)$ and $F_2(\cdot)$ such that*

$$
\begin{aligned}
|\tau(\mathbf{t}) - A(\mathbf{t} : 3)| &= \exp\left\{-\frac{1}{2}\mathbf{t}'\Omega\mathbf{t}\right\} \times F_1(\|\mathbf{t}\|) \cdot O(n^{-\frac{3}{2}}) \\
&\quad + \exp\{-\mathbf{t}'Q_0\mathbf{t}\} \times F_2(\|\mathbf{t}\|) \cdot O(n^{-\frac{3}{2}}),
\end{aligned} \tag{3.2.48}
$$

where

$$
\begin{aligned}
A(\mathbf{t} : 3) &= \exp\left\{-\frac{1}{2}\mathbf{t}'\Omega\mathbf{t}\right\} \times \left[1 + \frac{1}{\sqrt{n}}i\mu t_1 \right. \\
&\quad + \frac{1}{6\sqrt{n}}\sum_{j=1}^{3}\sum_{k=1}^{3}\sum_{l=1}^{3}A_{jkl}(it_j)(it_k)(it_l) + \frac{1}{2n}\sum_{j=1}^{3}\sum_{k=1}^{3}B_{jk}(it_j)(it_k) \\
&\quad + \frac{1}{2n}\mu^2(it_1)^2 + \frac{\mu(it_1)}{6n}\sum_{j=1}^{3}\sum_{k=1}^{3}\sum_{l=1}^{3}A_{jkl}(it_j)(it_k)(it_l) \\
&\quad + \frac{1}{24n}\sum_{j=1}^{3}\sum_{k=1}^{3}\sum_{l=1}^{3}\sum_{m=1}^{3}A_{jklm}(it_j)(it_k)(it_l)(it_m) \\
&\quad \left. + \frac{1}{72n}\sum_{j=1}^{3}\sum_{k=1}^{3}\sum_{l=1}^{3}\sum_{j'=1}^{3}\sum_{k'=1}^{3}\sum_{l'=1}^{3}A_{jkl}A_{j'k'l'}(it_j)(it_k)(it_l)(it_{j'})(it_{k'})(it_{l'}) \right].
\end{aligned}
$$

Proof. From (3.2.46) we have

$$
\begin{aligned}
\tau(\mathbf{t}) &= \exp\left\{-\frac{1}{2}\mathbf{t}'\Omega\mathbf{t}\right\} \times \exp\left[it_1\left\{\frac{\mu}{\sqrt{n}} + o(n^{-1})\right\}\right. \\
&\quad + \frac{1}{2n}\sum_{j=1}^{3}\sum_{k=1}^{3}B_{jk}(it_j)(it_k) + \frac{1}{6\sqrt{n}}\sum_{j=1}^{3}\sum_{k=1}^{3}\sum_{l=1}^{3}A_{jkl}(it_j)(it_k)(it_l) \\
&\quad \left. + \frac{1}{24n}\sum_{j=1}^{3}\sum_{k=1}^{3}\sum_{l=1}^{3}\sum_{m=1}^{3}A_{jklm}(it_j)(it_k)(it_l)(it_m) + F_3(\|\mathbf{t}\|)O(n^{-\frac{3}{2}})\right], \tag{3.2.49}
\end{aligned}
$$

where $F_3(\cdot)$ is a polynomial function. Applying the relation

$$
\left|e^z - 1 - z - \frac{z^2}{2}\right| \le \frac{|z|^3}{6}e^{|z|} \tag{3.2.50}
$$

to the second exponential in the right-hand side of (3.2.49) we have

$$
\begin{aligned}
|\tau(t) - A(t:3)| &= \exp\left\{-\frac{1}{2}t'\Omega t\right\} \cdot F_1(\|t\|) \cdot O(n^{-\frac{3}{2}}) \\
&\quad + O(n^{-\frac{3}{2}}) \cdot F_4(\|t\|) \cdot \exp\left\{-\frac{1}{2}t'\Omega t\right\} \\
&\quad \times \exp\left[\left|it_1\left\{\frac{\mu}{\sqrt{n}} + o(n^{-1})\right\} + \frac{1}{2n}\sum_j\sum_k\{B_{jk} + O(n^{-\frac{1}{2}})\}(it_j)(it_k)\right.\right. \\
&\quad + \frac{1}{6\sqrt{n}}\sum_j\sum_k\sum_l\{A_{jkl} + O(n^{-1})\}(it_j)(it_k)(it_l) \\
&\quad + \frac{1}{24n}\sum_j\sum_k\sum_l\sum_m\{A_{jklm} + O(n^{-1})\}(it_j)(it_k)(it_l)(it_m) \\
&\quad + \frac{1}{120n^{3/2}}\sum_j\sum_k\sum_l\sum_m\sum_r\{A_{jklmr} + O(n^{-1})\} \\
&\quad \left.\left. \times (it_j)(it_k)(it_l)(it_m)(it_r) + R_6\right|\right], \tag{3.2.51}
\end{aligned}
$$

where $F_4(\cdot)$ is a polynomial function. Let $\omega > 0$ be the smallest eigen value of Ω. Then for sufficiently large n, we can choose $\delta_1 > 0$ so that

$$
\frac{\omega}{4} - \frac{\delta_1}{6}\sum_{j=1}^3\sum_{k=1}^3\sum_{l=1}^3|A_{jkl}| - \frac{\delta_1^2}{24}\sum_{j=1}^3\sum_{k=1}^3\sum_{l=1}^3\sum_{m=1}^3|A_{jklm}|
$$
$$
- \frac{\delta_1^3}{120}\sum_{j=1}^3\sum_{k=1}^3\sum_{l=1}^3\sum_{m=1}^3\sum_{r=1}^3|A_{jklmr}| - 2^5\delta_1^4\sum_{j=1}^3\sum_{k=1}^3\sum_{l=1}^3\sum_{m=1}^3\sum_{r=1}^3\sum_{s=1}^3|A_{jklmrs}| > 0. \tag{3.2.52}
$$

Thus the last exponential term in (3.2.51) is dominated by

$$
\exp\delta_1\{|\mu| + o(n^{-\frac{1}{2}})\}\exp\|t\|^2\left\{\frac{\omega}{4} + O(n^{-1})\right\}, \quad \text{for} \quad \|t\| \le \delta_1\sqrt{n}, \tag{3.2.53}
$$

which implies the existence of Q_0 in (3.2.48).

We also have

Lemma 3.2.6. *Under Assumptions 3.2.1 – 3.2.5, for every $\eta > 0$, there exists $\delta_2 > 0$ such that*

$$
|\tau(t)| \le (1 + 4\delta_2\eta)^{-q(n)/4}, \tag{3.2.54}
$$

for all t satisfying $\|t\| \ge \eta\sqrt{n}$, where $q(n) = [cn]$, for some constant c.

Proof. Remembering that ρ_1 is the largest eigenvalue of S, we can see $\rho_1^2 = \max_e e'S^2e$, where $e = (e_1, \ldots, e_n)' \in R^n$ and $e'e = 1$. Then

$$
\begin{aligned}
e'S^2e &= e'\left\{\Sigma^{\frac{1}{2}}(t_1G + t_2\dot{G} + t_3\ddot{G})\Sigma^{\frac{1}{2}}\right\}^2e \\
&\le 2t_1^2e'\Sigma^{\frac{1}{2}}G\Sigma G\Sigma^{\frac{1}{2}}e + 2t_2^2e'\Sigma^{\frac{1}{2}}\dot{G}\Sigma\dot{G}\Sigma^{\frac{1}{2}}e + 2t_3^2e'\Sigma^{\frac{1}{2}}\ddot{G}\Sigma\ddot{G}\Sigma^{\frac{1}{2}}e. \tag{3.2.55}
\end{aligned}
$$

It is not difficult to show that

$$
e'\Sigma^{\frac{1}{2}}G\Sigma G\Sigma^{\frac{1}{2}}e \le c_1, \tag{3.2.56}
$$

$$\mathbf{e}'\Sigma^{\frac{1}{2}}\dot{G}\Sigma\dot{G}\Sigma^{\frac{1}{2}}\mathbf{e} \leq c_2, \tag{3.2.57}$$

$$\mathbf{e}'\Sigma^{\frac{1}{2}}\ddot{G}\Sigma\ddot{G}\Sigma^{\frac{1}{2}}\mathbf{e} \leq c_3, \tag{3.2.58}$$

where c_1, c_2 and c_3 are some positive constants. For exposition we prove (3.2.56). Since $f_\theta(\lambda)$, $h_\theta(\lambda) \in \mathbf{D}_d$ and $g_\theta \in \mathbf{D}_{ARMA}^c$, we can set

$$f_1 = \max_\lambda f_\theta(\lambda) < \infty,$$

$$h_1 = \max_\lambda |h_\theta(\lambda)| < \infty,$$

$$g_1 = \min_\lambda g_\theta(\lambda) > 0.$$

Using discussions of Anderson(1971, p.573–574) we obtain

$$\mathbf{e}'\Sigma^{\frac{1}{2}}G\Sigma G\Sigma^{\frac{1}{2}}\mathbf{e}$$

$$= \mathbf{e}'\Sigma^{\frac{1}{2}}A^{-1}BA^{-1}\Sigma A^{-1}BA^{-1}\Sigma^{\frac{1}{2}}\mathbf{e}$$

$$\leq \mathbf{e}'\Sigma^{\frac{1}{2}}A^{-1}BA^{-1}\begin{pmatrix} 2\pi f_1 & & 0 \\ & \ddots & \\ 0 & & 2\pi f_1 \end{pmatrix}A^{-1}BA^{-1}\Sigma^{\frac{1}{2}}\mathbf{e}$$

$$= 2\pi f_1 \mathbf{e}'\Sigma^{\frac{1}{2}}A^{-1}BA^{-\frac{1}{2}}A^{-1}A^{-\frac{1}{2}}BA^{-1}\Sigma^{\frac{1}{2}}\mathbf{e}$$

$$\leq 2\pi f_1 \mathbf{e}'\Sigma^{\frac{1}{2}}A^{-1}BA^{-\frac{1}{2}}\begin{pmatrix} 2\pi g_1 & & 0 \\ & \ddots & \\ 0 & & 2\pi g_1 \end{pmatrix}^{-1}A^{-\frac{1}{2}}BA^{-1}\Sigma^{\frac{1}{2}}\mathbf{e}$$

$$= \frac{f_1}{g_1}\mathbf{e}'\Sigma^{\frac{1}{2}}A^{-1}BA^{-1}BA^{-1}\Sigma^{\frac{1}{2}}\mathbf{e}$$

$$\leq \frac{f_1}{2\pi g_1^2}\mathbf{e}'\Sigma^{\frac{1}{2}}A^{-1}B^{\frac{1}{2}}BB^{\frac{1}{2}}A^{-1}\Sigma^{\frac{1}{2}}\mathbf{e}$$

$$\leq \frac{2\pi f_1 h_1^2}{g_1^2}\mathbf{e}'\Sigma^{\frac{1}{2}}A^{-1}A^{-1}\Sigma^{\frac{1}{2}}\mathbf{e}$$

$$\leq \frac{f_1^2 h_1^2}{g_1^4}\mathbf{e}'\mathbf{e} = \frac{f_1^2 h_1^2}{g_1^4} < \infty \tag{3.2.59}$$

which yields (3.2.56). The proofs of (3.2.57) and (3.2.58) are similar to the above. It follows from (3.2.55) that

$$\rho_n^2 \leq \cdots \leq \rho_1^2 \leq \|\mathbf{t}\|^2 \cdot d_{11} \tag{3.2.60}$$

for any $\mathbf{t}$, where d_{11} is a positive constant. While by Theorem 2.2.1, we get

$$n^{-1}\sum_{j=1}^{n}\rho_j^2 = n^{-1}\mathrm{tr}S^2$$

$$= \frac{1}{2\pi}\int_{-\pi}^{\pi}\{t_1 A(\lambda) + t_2 B(\lambda) + t_3 C(\lambda)\}^2\,d\lambda + \|\mathbf{t}\|^2 O(n^{-1})$$

$$= \frac{\|\mathbf{t}\|^2}{2\pi}\int_{-\pi}^{\pi}\left\{\frac{t_1 A(\lambda) + t_2 B(\lambda) + t_3 C(\lambda)}{\|\mathbf{t}\|}\right\}^2\,d\lambda + \|\mathbf{t}\|^2 O(n^{-1}), \tag{3.2.61}$$

where

$$A(\lambda) = \frac{1}{2}\frac{\partial}{\partial\theta}f_\theta(\lambda) \cdot f_\theta(\lambda)^{-1},$$

$$B(\lambda) = \frac{1}{2}\left\{-2f_\theta(\lambda)^{-2}\left(\frac{\partial}{\partial\theta}f_\theta(\lambda)\right)^2 + f_\theta(\lambda)^{-1}\frac{\partial^2}{\partial\theta^2}f_\theta(\lambda)\right\},$$

and

$$C(\lambda) = \frac{1}{2}\left\{6f_\theta(\lambda)^{-3}\left(\frac{\partial}{\partial\theta}f_\theta(\lambda)\right)^3 - 6f_\theta(\lambda)^{-2}\frac{\partial}{\partial\theta}f_\theta(\lambda) \cdot \frac{\partial^2}{\partial\theta^2}f_\theta(\lambda) + f_\theta(\lambda)^{-1}\frac{\partial^3}{\partial\theta^3}f_\theta(\lambda)\right\}.$$

Since we are now assuming that the matrix Ω is nonsingular, the functions $A(\lambda)$, $B(\lambda)$ and $C(\lambda)$ are linearly independent in the L_2-norm $(\int_{-\pi}^{\pi}|\cdot|^2\,d\lambda)$. Thus we can show that for sufficiently large n, there exists $d_{12} > 0$ such that

$$n^{-1}\sum_{j=1}^{n}\rho_j^2 \geq \|\mathbf{t}\|^2 d_{12}, \tag{3.2.62}$$

for any $\mathbf{t}$. The relations (3.2.60) and (3.2.62) imply that there exists $\delta_2 > 0$ and $q(n) = [cn]$ such that

$$\rho_1^2 \geq \ldots \geq \rho_{q(n)}^2 \geq \delta_2\|\mathbf{t}\|^2.$$

The proof of Lemma 3.2.6 follows from the inequalities;

$$\begin{aligned}|\tau(\mathbf{t})| &= \prod_{j=1}^{n}\left(1 + \frac{4}{n}\rho_j^2\right)^{-\frac{1}{4}}\\ &\leq \prod_{j=1}^{q(n)}\left(1 + \frac{4\delta_2}{4}\|\mathbf{t}\|^2\right)^{-\frac{1}{4}}\\ &\leq (1 + 4\delta_2\eta)^{-q(n)/4}, \quad\text{for } \|\mathbf{t}\| \geq \eta\sqrt{n}.\end{aligned}$$

Next let us consider the Edgeworth expansion for $\mathbf{Z}$. We set down $B(\mathbf{x} : r) = \{\mathbf{z} \in R^p : \|\mathbf{z} - \mathbf{x}\| \leq r, \ \mathbf{x} \in R^p\}$. The following lemma follows from Lemma 3.1.4 on replacing f by the indicator function of a Borel subset of R^p (see Bhattacharya and Rao(1976, p.97–98, 113)).

Lemma 3.2.7. *Let P and Q be probability measures on R^p and $\mathbf{B}^p$ the class of all Borel subsets of R^p. Let ϵ be a positive number. Then there exists a kernel probability measure K_ϵ such that*

$$\sup_{B\in\mathbf{B}^p}|P(B) - Q(B)| \leq \frac{2}{3}\|(P - Q)\star K_\epsilon\| + \frac{4}{3}\sup_{B\in\mathbf{B}^p}Q\{(\partial B)^{2\epsilon}\}, \tag{3.2.63}$$

where K_ϵ satisfies

$$K_\epsilon(B(0,r)^c) = O\left\{\left(\frac{\epsilon}{r}\right)^3\right\}, \tag{3.2.64}$$

and the Fourier transform $\hat{K}_\epsilon$ satisfies

$$\hat{K}_\epsilon(\mathbf{t}) = 0 \quad\text{for } \|\mathbf{t}\| \geq \frac{8p^{4/3}}{\pi^{1/3}\epsilon}.$$

For $B \in \mathbf{\dot{B}}^3$, define

$$
\begin{aligned}
Q_{\mathbf{Z}}^{(3)}(B) = \int_B N(\mathbf{z} : \Omega) &\left[1 + \frac{\mu}{\sqrt{n}} H_1(\mathbf{z}) + \frac{1}{6\sqrt{n}} \sum_{j=1}^{3} \sum_{k=1}^{3} \sum_{l=1}^{3} A_{jkl} H_{jkl}(\mathbf{z}) \right.\\
&+ \frac{\mu^2}{2n} H_{11}(\mathbf{z}) + \frac{1}{2n} \sum_{j=1}^{3} \sum_{k=1}^{3} B_{jk} H_{jk}(\mathbf{z}) + \frac{1}{6n} \sum_{j=1}^{3} \sum_{k=1}^{3} \sum_{l=1}^{3} \mu A_{jkl} H_{1jkl}(\mathbf{z})\\
&+ \frac{1}{24n} \sum_{j=1}^{3} \sum_{k=1}^{3} \sum_{l=1}^{3} \sum_{m=1}^{3} A_{jklm} H_{jklm}(\mathbf{z})\\
&\left. + \frac{1}{72n} \sum_{j=1}^{3} \sum_{k=1}^{3} \sum_{l=1}^{3} \sum_{j'=1}^{3} \sum_{k'=1}^{3} \sum_{l'=1}^{3} A_{jkl} A_{j'k'l'} H_{jklj'k'l'}(\mathbf{z}) \right] d\mathbf{z} \qquad (3.2.65)
\end{aligned}
$$

where $\mathbf{z} = (z_1, z_2, z_3)'$,

$$
N(\mathbf{z} : \Omega) = (2\pi)^{-\frac{3}{2}} |\Omega|^{-\frac{1}{2}} \exp{-\frac{1}{2} \mathbf{z}' \Omega^{-1} \mathbf{z}},
$$

$$
H_{j_1 \ldots j_s}(\mathbf{z}) = \frac{(-1)^s}{N(\mathbf{z} : \Omega)} \cdot \frac{\partial^s}{\partial z_{j_1} \ldots \partial z_{j_s}} N(\mathbf{z} : \Omega).
$$

This measure $Q_{\mathbf{Z}}^{(3)}(\cdot)$ corresponds to the characteristic function $A(\mathbf{t} : 3)$ in Lemma 3.2.5. Then we have

Lemma 3.2.8. *Suppose that Assumptions 3.2.1 – 3.2.5 are satisfied. Then*

$$
\sup_{B \in \mathbf{B}^3} \left| P_\theta^n \{ \mathbf{Z} \in B \} - Q_{\mathbf{Z}}^{(3)}(B) \right| = o(n^{-1}) + \frac{4}{3} \sup_{B \in \mathbf{B}^3} Q_{\mathbf{Z}}^{(3)} \{ (\partial B)^{2\epsilon} \}, \qquad (3.2.66)
$$

uniformly for $\theta \in \Theta$, where $\epsilon = n^{-1-\rho}$, $0 < \rho < \frac{1}{2}$.

Proof. Substituting $P_\theta^n \{ \mathbf{Z} \in B \}$ and $Q_{\mathbf{Z}}^{(3)}(B)$ for $P(B)$ and $Q(B)$ in Lemma 3.2.7, respectively, we have

$$
\sup_{B \in \mathbf{B}^3} \left| P_\theta^n \{ \mathbf{Z} \in B \} - Q_{\mathbf{Z}}^{(3)}(B) \right| \le \frac{2}{3} \left\| (P_\theta^n - Q_{\mathbf{Z}}^{(3)}) \star K_\epsilon \right\| + \frac{4}{3} \sup_{B \in \mathbf{B}^3} Q_{\mathbf{Z}}^{(3)} \{ (\partial B)^{2\epsilon} \}. \qquad (3.2.67)
$$

Note that

$$
\begin{aligned}
&\left\| (P_\theta^n - Q_{\mathbf{Z}}^{(3)}) \star K_\epsilon \right\|\\
&= 2 \sup \left\{ \left| (P_\theta^n - Q_{\mathbf{Z}}^{(3)}) \star K_\epsilon(B) \right|; B \in \mathbf{B}^3 \right\}\\
&\le 2 \sup \left\{ \left| (P_\theta^n - Q_{\mathbf{Z}}^{(3)}) \star K_\epsilon(B) \right|; B \in \mathbf{B}^3 \text{ and } B \subset B(0, r_n) \right\}\\
&\quad + 2 \sup \left\{ \left| (P_\theta^n - Q_{\mathbf{Z}}^{(3)}) \star K_\epsilon(B) \right|; B \in \mathbf{B}^3 \text{ and } B \subset B(0, r_n)^c \right\}, \qquad (3.2.68)
\end{aligned}
$$

where $r_n = n^\tau$, $0 < \tau < \frac{1}{6}$. Then, for $B \subset B(0, r_n)^c$ we have

$$
\begin{aligned}
\left| (P_\theta^n - Q_{\mathbf{Z}}^{(3)}) \star K_\epsilon(B) \right| &\le |P_\theta^n \star K_\epsilon(B)| + \left| Q_{\mathbf{Z}}^{(3)} \star K_\epsilon(B) \right|\\
&\le P_\theta^n \left(\|\mathbf{Z}\| \ge \frac{r_n}{2} \right) + K_\epsilon \left\{ B \left(0, \frac{r_n}{2} \right)^c \right\}\\
&\quad + Q_{\mathbf{Z}}^{(3)} \left\{ B \left(0, \frac{r_n}{2} \right)^c \right\} + K_\epsilon \left\{ B \left(0, \frac{r_n}{2} \right)^c \right\}. \qquad (3.2.69)
\end{aligned}
$$

It is easy to check

$$Q_{\mathbf{Z}}^{(3)} \left\{ B \left(\mathbf{0}, \frac{r_n}{2} \right)^c \right\} = o(n^{-1}). \tag{3.2.70}$$

The relations (3.2.24), (3.2.25) and (3.2.26) imply

$$P_{\theta}^n \left(\|\mathbf{Z}\| \geq \frac{r_n}{2} \right) = o(n^{-1}). \tag{3.2.71}$$

It follows from (3.2.64) that

$$K_{\epsilon} \left\{ B \left(\mathbf{0}, \frac{r_n}{2} \right)^c \right\} = O(n^{-3-3\rho-3\tau}) = o(n^{-1}). \tag{3.2.72}$$

Thus we have

$$\left| \left(P_{\theta}^n - Q_{\mathbf{Z}}^{(3)} \right) \star K_{\epsilon}(B) \right| = o(n^{-1}), \quad \text{for} \quad B \subset B(\mathbf{0}, r_n)^c. \tag{3.2.73}$$

Now we have only to evaluate

$$\sup \left\{ \left| \left(P_{\theta}^n - Q_{\mathbf{Z}}^{(3)} \right) \star K_{\epsilon}(B) \right| ; B \subset B(\mathbf{0}, r_n) \right\}.$$

By Fourier inversion we have

$$\left| \left(P_{\theta}^n - Q_{\mathbf{Z}}^{(3)} \right) \star K_{\epsilon}(B) \right| \leq (2\pi)^{-3} \frac{2\pi^{3/2} r_n^3}{3\Gamma(3/2)} \int \left| \left(\hat{P}_{\theta}^n - \hat{Q}_{\mathbf{Z}}^{(3)} \right)(\mathbf{t}) \hat{K}_{\epsilon}(\mathbf{t}) \right| d\mathbf{t}, \tag{3.2.74}$$

where $\hat{P}_{\theta}^n$ and $\hat{Q}_{\mathbf{Z}}^{(3)}$ are the Fourier transforms of P_{θ}^n and $Q_{\mathbf{Z}}^{(3)}$, respectively. By Lemma 3.2.5 and noting $\hat{Q}_{\mathbf{Z}}^{(3)}(\mathbf{t}) = A(\mathbf{t} : 3)$, the right-hand side of (3.2.74) is dominated by

$$O(n^{3\tau - \frac{3}{2}}) \int_{\|\mathbf{t}\| \leq \delta_1 \sqrt{n}} \left| \exp \left\{ -\frac{1}{2} \mathbf{t}' \Omega \mathbf{t} \right\} \times F_1(\|\mathbf{t}\|) \right.$$
$$+ \exp \{ -\mathbf{t}' Q_0 \mathbf{t} \} \times F_2(\|\mathbf{t}\|) \Big| \left| \hat{K}_{\epsilon}(\mathbf{t}) \right| d\mathbf{t}$$
$$+ O(n^{3\tau}) \int_{\delta_1 \sqrt{n} < \|\mathbf{t}\| \leq 8(3)^{4/3} n^{1+\rho}/\pi^{1/3}} \left| \left(\hat{P}_{\theta}^n - \hat{Q}_{\mathbf{Z}}^{(3)} \right) \star \hat{K}_{\epsilon}(\mathbf{t}) \right| d\mathbf{t}. \tag{3.2.75}$$

Evidently the first term of the above is of order $o(n^{-1})$. In view of Lemma 3.2.6, the second term of (3.2.75) is dominated by

$$O(n^{3\tau}) \int_{\delta_1 \sqrt{n} < \|\mathbf{t}\| \leq 8(3)^{4/3} n^{1+\rho}/\pi^{1/3}} \left\{ \left| \hat{P}_{\theta}^n(\mathbf{t}) \right| + \left| \hat{Q}_{\mathbf{Z}}^{(3)}(\mathbf{t}) \right| \right\} d\mathbf{t}$$
$$\leq O(n^{3\tau}) \int_{\delta_1 \sqrt{n} < \|\mathbf{t}\| \leq d_{13} n^{1+\rho}} (1 + 4\delta_2 \delta_1)^{-q(n)/4} d\mathbf{t} + o(n^{-1}), \tag{3.2.76}$$

where d_{13} is an appropriate positive constant. The above (3.2.76) is of order

$$O(n^{3\tau + 3 + 3\rho})(1 + 4\delta_2 \delta_1)^{-q(n)/4} + o(n^{-1}) = o(n^{-1}).$$

Therefore we have proved

$$\sup \left\{ \left| \left(P_{\theta}^n - Q_{\mathbf{Z}}^{(3)} \right) \star K_{\epsilon}(B) \right| ; B \subset B(\mathbf{0}, r_n) \right\} = o(n^{-1}),$$

which completes the proof.

Now we proceed to the proof for (2) of Theorem 3.2.1.

Proof of (2) in Theorem 3.2.1. Consider the transformation

$$
\begin{aligned}
W_1(\theta) &= Z_1(\theta), \\
W_2(\theta) &= Z_2(\theta) - J(\theta)I(\theta)^{-1}Z_1(\theta), \\
W_3(\theta) &= Z_3(\theta) - L(\theta)I(\theta)^{-1}Z_1(\theta).
\end{aligned}
\tag{3.2.77}
$$

Henceforth, for simplicity we sometimes use W_1, W_2 and W_3 instead of $W_1(\theta)$, $W_2(\theta)$ and $W_3(\theta)$, respectively. Evidently (3.2.77) is a continuous bijective transformation. We denote (3.2.77) by $\mathbf{W} = \chi(\mathbf{Z})$, where $\mathbf{W} = (W_1, W_2, W_3)'$. By Lemma 3.2.8, we have

$$
\sup_{B \in \mathbf{B}^3} \left| P_\theta^n \{ \mathbf{Z} \in \chi^{-1}(B)\} - Q_{\mathbf{Z}}^{(3)}\{\chi^{-1}(B)\} \right| = \frac{4}{3} \sup_{B \in \mathbf{B}^3} Q_{\mathbf{Z}}^{(3)} \left[\{\partial \chi^{-1}(B)\}^{2\epsilon} \right] + o(n^{-1}).
\tag{3.2.78}
$$

Here we put $Q_{\mathbf{W}}^{(3)}(B) = Q_{\mathbf{Z}}^{(3)}\{\chi^{-1}(B)\}$. Then it is not difficult to show

$$
\begin{aligned}
Q_{\mathbf{W}}^{(3)}(B) &= \int_B N(\omega_1 : I) N(\omega_2, \omega_3 : \Omega_2) \\
&\quad \times \Bigg[1 + \frac{1}{\sqrt{n}} \sum_{j=1}^{3} C_j^{(1)} H_j(\boldsymbol{\omega}) + \frac{1}{6\sqrt{n}} \sum_{j=1}^{3} \sum_{k=1}^{3} \sum_{l=1}^{3} C_{jkl}^{(1)} H_{jkl}(\boldsymbol{\omega}) \\
&\quad + \frac{1}{2n} \sum_{j=1}^{3} \sum_{k=1}^{3} \left(C_{jk}^{(3)} + C_j^{(1)} C_k^{(1)} \right) H_{jk}(\boldsymbol{\omega}) \\
&\quad + \frac{1}{n} \sum_{j=1}^{3} \sum_{k=1}^{3} \sum_{l=1}^{3} \sum_{m=1}^{3} \left(\frac{C_j^{(1)} C_{klm}^{(1)}}{6} + \frac{C_{jklm}^{(1)}}{24} \right) H_{jklm}(\boldsymbol{\omega}) \\
&\quad + \frac{1}{72n} \sum_{j=1}^{3} \sum_{k=1}^{3} \sum_{l=1}^{3} \sum_{j'=1}^{3} \sum_{k'=1}^{3} \sum_{l'=1}^{3} C_{jkl}^{(1)} C_{j'k'l'}^{(1)} H_{jklj'k'l'}(\boldsymbol{\omega}) \Bigg] \, d\boldsymbol{\omega} \\
&= \int_B q_n(\boldsymbol{\omega}) \, d\boldsymbol{\omega}, \quad \text{(say)},
\end{aligned}
\tag{3.2.79}
$$

where $\boldsymbol{\omega} = (\omega_1, \omega_2, \omega_3)'$,

$$
N(\omega_1 : I) = (2\pi)^{-\frac{1}{2}} I^{-\frac{1}{2}} \exp -\frac{\omega_1^2}{2I},
$$

$$
N(\omega_2, \omega_3 : \Omega_2) = (2\pi)^{-1} |\Omega_2|^{-\frac{1}{2}} \exp -\frac{1}{2}(\omega_2, \omega_3)\Omega_2^{-1} \begin{pmatrix} \omega_2 \\ \omega_3 \end{pmatrix},
$$

$$
\Omega_2 = \begin{pmatrix} \Omega_{22} & \Omega_{23} \\ \Omega_{32} & \Omega_{33} \end{pmatrix}, \quad 2 \times 2\text{-matrix}.
$$

For examples we can see that

$$
C_1^{(1)} = \mu(\theta), \quad C_2^{(1)} = -\frac{J(\theta)\mu(\theta)}{I(\theta)}, \quad C_3^{(1)} = -\frac{L(\theta)\mu(\theta)}{I(\theta)},
$$

$$
\Omega_{22} = M(\theta) - J(\theta)^2 I(\theta)^{-1}, \quad C_{112}^{(1)} = N(\theta) - \frac{J(\theta)K(\theta)}{I(\theta)}, \quad C_{1111}^{(1)} = H(\theta), \quad \text{etc.}
$$

Since χ is continuous we have

$$\partial \chi^{-1}(B) \subset \chi^{-1}(\partial B),$$

$$\{\partial \chi^{-1}(B)\}^{2\epsilon} \subset \{\chi^{-1}(\partial B)\}^{2\epsilon}, \qquad (3.2.80)$$

and there exists $a > 0$ such that

$$\{\chi^{-1}(\partial B)\}^{2\epsilon} \subset \chi^{-1}\{(\partial B)^{a\epsilon}\}. \qquad (3.2.81)$$

Thus we have

Lemma 3.2.9. *Under Assumptions 3.2.1 – 3.2.5,*

$$\sup_{B \in \mathbf{B}^3} \left| P_\theta^n(\mathbf{W} \in B) - Q_{\mathbf{W}}^{(3)}(B) \right| = \frac{4}{3} \sup_{B \in \mathbf{B}^3} Q_{\mathbf{W}}^{(3)}\{(\partial B)^{a\epsilon}\} + o(n^{-1}), \qquad (3.2.82)$$

uniformly for $\theta \in \Theta$, where a is a positive constant and $\epsilon = n^{-1-\rho}$, $0 < \rho < \frac{1}{2}$.

Now we rewrite $U_n(\theta)$ in Lemma 3.2.4 as

$$
\begin{aligned}
U_n(\mathbf{W}) \;=\; & \frac{W_1}{I_n} + \frac{W_1 W_2}{I^2 \sqrt{n}} - \frac{J+K}{2I^3 \sqrt{n}} W_1^2 + \frac{1}{I^3 n}\left\{ W_1 W_2^2 - \frac{5J+3K}{2I} W_1^2 W_2 \right. \\
& \left. + \frac{1}{2} W_1^2 W_3 + \frac{2J^2 + 3KJ + K^2}{2I^2} W_1^3 - \frac{L + 3M + 6N + H}{6I} W_1^3 \right\}. \qquad (3.2.83)
\end{aligned}
$$

Consider the following transformation

$$
\begin{aligned}
S_1 &= U_n(\mathbf{W}), \\
S_2 &= W_2, \qquad (3.2.84) \\
S_3 &= W_3.
\end{aligned}
$$

We denote (3.2.84) by $\mathbf{S} = \Psi(\mathbf{W})$, where $\mathbf{S} = (S_1, S_2, S_3)'$. For sufficiently large n, we can take a set

$$M_n = \left\{ \mathbf{W} : |W_i| \le c_i n^\alpha, \ 0 < \alpha < \frac{1}{8}, \ i = 1, 2, 3 \right\}$$

such that Ψ is a C^∞-mapping on M_n. From Lemma 3.2.9,

$$
\begin{aligned}
\sup_{B \in \mathbf{B}} &\left| P_\theta^n \left\{ \mathbf{W} \in \Psi^{-1}(B \times R^2) \right\} - Q_{\mathbf{W}}^{(3)}\left\{ \Psi^{-1}(B \times R^2) \right\} \right| \\
&= \frac{4}{3} \sup_{B \in \mathbf{B}} Q_{\mathbf{W}}^{(3)}\left\{ \left(\partial \Psi^{-1}(B \times R^2) \right)^{a\epsilon} \right\} + o(n^{-1}). \qquad (3.2.85)
\end{aligned}
$$

We can see that

$$
\begin{aligned}
Q_{\mathbf{W}}^{(3)}\left\{ \Psi^{-1}(B \times R^2) \right\} &= \int_{\Psi^{-1}(B \times R^2)} q_n(\boldsymbol{\omega}) \, d\boldsymbol{\omega} \\
&= \int_{M_n \cap \Psi^{-1}(B \times R^2)} q_n(\boldsymbol{\omega}) \, d\boldsymbol{\omega} + o(n^{-1}) \\
&= \int_{N_n \cap \{B \times R^2\}} q_n\left\{ \Psi^{-1}(\mathbf{S}) \right\} |J| \, d\mathbf{S} + o(n^{-1}), \qquad (3.2.86)
\end{aligned}
$$

where $N_n = \Psi(M_n)$ and $|J|$ is the Jacobian. Since we can solve so that

$$
\begin{aligned}
W_1 &= I_n S_1 - \frac{1}{\sqrt{n}} S_1 S_2 + \frac{J+K}{2\sqrt{n}} S_1^2 + \frac{J}{In} S_1^2 S_2 - \frac{1}{2n} S_1^2 S_3 \\
&\quad - \frac{J^2 + JK}{2In} S_1^3 + \frac{L + 3M + 6N + H}{6n} S_1^3 + o(n^{-1}),
\end{aligned}
\tag{3.2.87}
$$

uniformly on M_n, it is not difficult to show that

$$
q_n\{\Psi^{-1}(\mathbf{S})\}|J| = N(I_n S_1)N(S_2, S_3 : \Omega_2)\left\{ I_n + \frac{1}{\sqrt{n}} p_1(\mathbf{S}) + \frac{1}{n} p_2(\mathbf{S}) + o(n^{-1}) \right\},
\tag{3.2.88}
$$

uniformly on N_n, where $p_1(\mathbf{S})$ and $p_2(\mathbf{S})$ are polynomials of $\mathbf{S}$. Thus we have

$$
\begin{aligned}
Q_{\mathbf{W}}^{(3)}\left\{ \Psi^{-1}(B \times R^2) \right\} &= \int_{\{B \times R^2\} \cap N_n} N(I_n S_1)N(S_2, S_3 : \Omega) \\
&\quad \times \left\{ I_n + \frac{1}{\sqrt{n}} p_1(\mathbf{S}) + \frac{1}{n} p_2(\mathbf{S}) + o(n^{-1}) \right\} d\mathbf{S} + o(n^{-1}) \\
&= \int_B N(I_n S_1)\left[\iint_{R^2} \left\{ I_n + \frac{1}{\sqrt{n}} P_1(\mathbf{S}) + \frac{1}{n} P_2(\mathbf{S}) \right\} \right. \\
&\quad \left. \times N(S_2, S_3 : \Omega_2)\, dS_2 dS_3 \right] dS_1 + o(n^{-1}).
\end{aligned}
\tag{3.2.89}
$$

Calculating the square bracket in (3.2.89), and noting that

$$
Q_{\mathbf{W}}^{(3)}\left\{ \left(\partial \Psi^{-1}(B \times R^2)\right)^{a\epsilon} \right\} \le Q_{\mathbf{W}}^{(3)}\left[\Psi^{-1}\left\{ (\partial B)^{b\epsilon} \times R^2 \right\} \right] \quad \text{for some } b > 0,
$$

we have

$$
\begin{aligned}
\sup_{B \in \mathbf{B}_0} & \left| P_\theta^n\left\{ \sqrt{nI}\left(\hat{\theta}_n - \theta\right) \in B \right\} - \int_B \phi(x) p_3^n(x)\, dx \right| \\
&= \frac{4}{3} \sup_{B \in \mathbf{B}_0} \int_{(\partial B)^{b\epsilon}} \phi(x) p_3^n(x)\, dx + o(n^{-1})
\end{aligned}
\tag{3.2.90}
$$

(remember (3.2.85)). Here

$$
\begin{aligned}
p_3^n(x) &= 1 + \frac{\alpha_1}{\sqrt{n}} x + \frac{\gamma_1}{6\sqrt{n}}(x^3 - 3x) + \frac{\rho_2 + \alpha_1^2}{2n}(x^2 - 1) \\
&\quad + \frac{1}{n}\left(\frac{\delta_1}{24} + \frac{\alpha_1 \gamma_1}{6} \right)(x^4 - 6x^2 + 3) + \frac{\gamma_1^2}{72n}(x^6 - 15x^4 + 45x^2 - 15),
\end{aligned}
\tag{3.2.91}
$$

where

$$
\begin{aligned}
\alpha_1 &= -\frac{J+K}{2I^{3/2}} + \frac{\mu}{I^{1/2}}, \\
\gamma_1 &= -\frac{3J + 2K}{I^{3/2}}, \\
\rho_2 &= \frac{2\eta}{I} + \frac{\Delta}{I} + \frac{7J^2 + 14JK + 5K^2}{2I^3} - \frac{L + 4N + H}{I^2} - \frac{2\mu(2J + K)}{I^2}, \\
\delta_1 &= \frac{12(2J + K)(J + K)}{I^3} - \frac{4L + 12N + 3H}{I^2},
\end{aligned}
$$

where

$$\text{Var}\{Z_1(\theta)\} = I(\theta) + \frac{\Delta(\theta)}{n} + o(n^{-1}), \tag{3.2.92}$$

$$I_n(\theta) = I(\theta) - \frac{\eta(\theta)}{n} + o(n^{-1}). \tag{3.2.93}$$

Remembering (3.2.6) we have proved (2) of Theorem 3.2.1. More explicit forms of (3.2.91) for the exact maximum likelihood estimators are given in Section 2.3.

Now we turn to discuss third-order asymptotic properties of generalized maximum likelihood estimators in the class of third-order AMU estimators. In general the generalized maximum likelihood estimator $\hat{\theta}_n$ is not third-order AMU. To be so a modification of $\hat{\theta}_n$ is required. The following theorem gives the validity of Edgeworth expansion for modified estimators of $\hat{\theta}_n$.

Theorem 3.2.2. *Suppose that $m(\theta)$ is a continuously twice differentiable function. Define*

$$\hat{\theta}_m = \hat{\theta}_n + \frac{1}{n}m(\hat{\theta}_n),$$

then

$$\sup_{B \in \mathbf{B}_0} \left| P_\theta^n \left[\sqrt{nI} \left(\hat{\theta}_m - \theta \right) \in B \right] - \int_B \phi(y) q_3^n(y) \, dy \right| = o(n^{-1}), \tag{3.2.94}$$

uniformly for $\theta \in C$, where

$$
\begin{aligned}
q_3^n(y) &= 1 + \frac{1}{\sqrt{n}} \left\{ \alpha_1 + \sqrt{I}m(\theta) \right\} y + \frac{\gamma_1}{6\sqrt{n}}(y^3 - 3y) \\
&\quad + \frac{1}{2n} \left\{ \rho_2 + \alpha_1^2 + 2m'(\theta) + Im(\theta)^2 + 2\alpha_1\sqrt{I}m(\theta) \right\} (y^2 - 1) \\
&\quad + \frac{1}{n} \left\{ \frac{\delta_1}{24} + \frac{\alpha_1\gamma_1}{6} + \frac{\gamma_1\sqrt{I}m(\theta)}{6} \right\} (y^4 - 6y^2 + 3) \\
&\quad + \frac{\gamma_1^2}{72n}(y^6 - 15y^4 + 45y^2 - 15).
\end{aligned}
$$

Proof. Since $m(\cdot)$ is continuously twice differentiable we have

$$
\begin{aligned}
\sqrt{nI}(\hat{\theta}_m - \theta) &= \sqrt{nI}(\hat{\theta}_n - \theta) + \frac{\sqrt{I}}{\sqrt{n}}m(\theta) + \sqrt{nI}(\hat{\theta}_n - \theta) \cdot \frac{m'(\theta)}{n} \\
&\quad + n^{-3/2} \left\{ \sqrt{nI}(\hat{\theta}_n - \theta) \right\}^2 m''(\theta\star) \Big/ \sqrt{I}, \tag{3.2.95}
\end{aligned}
$$

where $\theta \lessgtr \theta \star \lessgtr \hat{\theta}_n$. By (1) of Theorem 3.2.1 we have

$$P_\theta^n \left[\left\{ \sqrt{nI}(\hat{\theta}_n - \theta) \right\}^2 |m''(\theta\star)| \Big/ \sqrt{I} > n^{2\alpha} \right] = o(n^{-1}), \tag{3.2.96}$$

for $0 < \alpha < \frac{1}{4}$. Putting $\rho_n = n^{2\alpha - 1/2}$ in Lemma 3.2.3, we have only to derive the Edgeworth expansion for $a_n U_n + s_n$, where $a_n = \{1 + m'(\theta)/n\}$, $U_n = \sqrt{nI}(\hat{\theta}_n - \theta)$ and $s_n = \sqrt{I}m(\theta)/\sqrt{n}$. From Theorem 3.2.1,

$$\sup_{B \in \mathbf{B}_0} \left| P_\theta^n \{ U_n \in B \} - \int_B \phi(x) p_3^n(x) \, dx \right| = o(n^{-1}). \tag{3.2.97}$$

Lemma 3.2.3 implies that

$$\sup_{B \in \mathbf{B}_0} \left| P_\theta^n \left\{ \sqrt{nI}(\hat{\theta}_m - \theta) \in B \right\} - P_\theta^n \{a_n U_n + s_n \in B\} \right| = o(n^{-1}). \quad (3.2.98)$$

Also we have

$$\sup_{B \in \mathbf{B}_0} \left| P_\theta^n \{a_n U_n + s_n \in B\} - \int_{a_n x + s_n \in B} \phi(x) p_3^n(x) \, dx \right| = o(n^{-1}). \quad (3.2.99)$$

Transformation $y = a_n x + s_n$ yields

$$\int_{a_n x + s_n \in B} \phi(x) p_3^n(x) \, dx = \int_B \phi(y) q_3^n(y) \, dy + o(n^{-1}). \quad (3.2.100)$$

The relations (3.2.98), (3.2.99) and (3.2.100) imply our assertion.

For $m(\theta) = K(\theta)/\{6I(\theta)^2\} - \mu(\theta)/I(\theta)$, we denote $\hat{\theta}_n^* = \hat{\theta}_m$. In this case we have

Corollary 3.2.1.

$$\sup_{B \in \mathbf{B}_0} \left| P_\theta^n \left\{ \sqrt{nI}(\hat{\theta}_n^* - \theta) \in B \right\} - \int_B \phi(x) \left[1 + \frac{\gamma_1}{6\sqrt{n}} x + \frac{\gamma_1}{6\sqrt{n}} (x^3 - 3x) \right. \right.$$

$$+ \frac{1}{2n} \left\{ \frac{2\eta}{I} + \frac{\Delta}{I} - \frac{2\mu'}{I} + \frac{135J^2 + 216JK + 70K^2}{36I^3} - \frac{3L + 9N + 2H}{3I^2} \right\} (x^2 - 1)$$

$$+ \frac{1}{n} \left\{ \frac{\delta_1}{24} + \frac{\gamma_1^2}{36} \right\} (x^4 - 6x^2 + 3)$$

$$\left. \left. + \frac{\gamma_1^2}{72n} (x^6 - 15x^4 + 45x^2 - 15) \right] \, dx \right| = o(n^{-1}). \quad (3.2.101)$$

Remark 3.2.2. Of course $\hat{\theta}_n^*$ belongs to $\mathbf{A}_2$, and we can see that the asymptotic distribution of $\hat{\theta}_n^*$ (up to second order) coincides with that of the second-order asymptotically efficient estimator (see Theorem 2.2.2).

Remark 3.2.3. It is easy to check that $\hat{\theta}_n^*$ is third-order AMU. Also it is $e_d = 2\eta + \Delta - 2\mu'$ that depends on the generalized maximum likelihood estimator.

Let $\hat{\theta}_i^*$ $(i = 1, 2)$ be the modified generalized maximum likelihood estimators with μ_i, Δ_i, η_i and m_i $(i = 1, 2)$ in place of μ, Δ, η and m, respectively. Then we have

Corollary 3.2.2. For $B = (-a, a)$, $a > 0$,

$$\lim_{n \to \infty} n \left[P_\theta^n \left\{ \sqrt{nI} \left(\hat{\theta}_1^* - \theta \right) \in B \right\} - P_\theta^n \left\{ \sqrt{nI} \left(\hat{\theta}_2^* - \theta \right) \in B \right\} \right]$$

$$= I^{-1} a \phi(a) \left\{ 2(\eta_2 - \eta_1) + \Delta_2 - \Delta_1 + 2(\mu_1' - \mu_2') \right\}. \quad (3.2.102)$$

Thus if $2\eta_1 + \Delta_1 - 2\mu_1'$ is smaller than $2\eta_2 + \Delta_2 - 2\mu_2'$, then $\hat{\theta}_1^*$ is better than $\hat{\theta}_2^*$ in third-order sense.

Example 3.2.3. Let $\{X_t\}$ be a Gaussian autoregressive process with the spectral density

$$f_\theta(\lambda) = \frac{\sigma^2}{2\pi} \frac{1}{|1 - \theta e^{i\lambda}|^2},$$

where $|\theta| < 1$.

Let $\hat{\theta}_1^*$ be the modified maximum likelihood estimator of θ (defined in Example 3.2.1). Also let $\hat{\theta}_2^*$ be the modified quasi-maximum likelihood estimator of θ (defined in Example 3.2.2). Then we can show that

$$\mu_1 = 0, \quad \Delta_1 = \frac{3\theta^2 - 1}{(1 - \theta^2)^2}, \quad \eta_1 = -\Delta_1,$$

$$\mu_2 = \frac{-\theta}{1 - \theta^2}, \quad \Delta_2 = \frac{-\theta^2 - 1}{(1 - \theta^2)^2}, \quad \eta_2 = 0.$$

For this case, the right-hand side of (3.2.102) is equal to

$$\frac{4}{I} a\phi(a) \frac{\theta^2}{1 - \theta^2} \geq 0,$$

which coincides with the results of Fujikoshi and Ochi(1984). That is, $\hat{\theta}_1^*$ is always better than $\hat{\theta}_2^*$.

CHAPTER 4

HIGHER ORDER ASYMPTOTIC SUFFICIENCY, ASYMPTOTIC ANCILLARITY IN TIME SERIES ANALYSIS

4.1. Higher order asymptotic sufficiency for Gaussian ARMA processes

Let $\mathbf{X}_n = (X_1, \ldots, X_n)'$ be an n-dimensional random vector which has an unknown parameter $\theta \in \Theta \subset R^1$. For the case when each X_j is independently and identically distributed, several works discussed the asymptotic sufficiency along the following directions. For an appropriate estimator $\hat{\theta}_n$ we define $t_n = \hat{\theta}_n + \{nF(\hat{\theta}_n)\}^{-1} \cdot L_n^{(1)}(\mathbf{X}_n, \hat{\theta}_n)$, where $F(\theta)$ is the Fisher information and $L_n^{(1)}(\mathbf{X}_n, \theta)$ is the derivative of the log-likelihood with respect to θ. Then LeCam(1956) showed that t_n is asymptotically sufficient in the sense that t_n is sufficient for a family $\{Q_{n,\theta}; \theta \in \Theta\}$ of probability measures and that $\|P_{n,\theta} - Q_{n,\theta}\| = o(1)$, uniformly on any compact set of Θ, where $\|\cdot\|$ is the variation norm. Under slight different conditions from those of LeCam, Pfanzagl(1972) showed that $\|P_{n,\theta} - Q_{n,\theta}\| = O(n^{-\frac{1}{2}})$, uniformly on every compact set of Θ, which improves the above result. Let $s_n = \{\hat{\theta}_n, L_n^{(1)}(\mathbf{X}_n, \hat{\theta}_n), \ldots, L_n^{(k)}(\mathbf{X}_n, \hat{\theta}_n)$, where $L_n^{(j)}(\mathbf{X}_n, \theta)$ is the jth derivative of $L_n(\mathbf{X}_n, \theta)$ with respect to θ. Suzuki(1978) proved that s_n is asymptotically sufficient up to order $o(n^{-(k-1)/2})$ in the sense that s_n is sufficient for a family $\{Q_{n,\theta}; \theta \in \Theta\}$ of probability measures and that

$$\|P_{n,\theta} - Q_{n,\theta}\| = o(n^{-\frac{k-1}{2}}),$$

uniformly on any compact set of Θ. Using the maximum likelihood estimator $\hat{\theta}_{ML}^{(n)}$ instead of $\hat{\theta}_n$, Ghosh and Subramanyam(1974) mentioned that $(\hat{\theta}_{ML}^{(n)}, L_n^{(2)}(\mathbf{X}_n, \hat{\theta}_{ML}^{(n)}), L_n^{(3)}(\mathbf{X}_n, \hat{\theta}_{ML}^{(n)}), L_n^{(4)}(\mathbf{X}_n, \hat{\theta}_{ML}^{(n)}))$ is asymptotically sufficient up to order $o(n^{-1})$.

In this section we shall extend these results to a Gaussian stationary process, i.e., $\mathbf{X}_n = (X_1, \ldots, X_n)'$ is a realization from a Gaussian ARMA process with the spectral density $f_\theta(\lambda)$. Then we show that $t_n = (\hat{\theta}_{GML}^{(n)}, L_n^{(1)}(\mathbf{X}_n, \hat{\theta}_{GML}^{(n)}), \ldots, L_n^{(4)}(\mathbf{X}_n, \hat{\theta}_{GML}^{(n)}))$ is asymptotically sufficient up to order $o(n^{-1})$, where $\hat{\theta}_{GML}^{(n)}$ is the generalized maximum likelihood estimator of θ defined in Section 3.2. This has the following application. Suppose that an estimator $\hat{\theta}_n$ is asymptotically unbiased up to order $o(n^{-1})$, i.e., $E_\theta(\hat{\theta}_n) - \theta = o(n^{-1})$, and let $W(\theta, \cdot)$ be an appropriate loss function. Then we can show that there exists an estimator $\hat{\Psi}(t_n)$ depending only on t_n such that

$$E_\theta W(\theta, \hat{\Psi}(t_n)) \le E_\theta W(\theta, \hat{\theta}_n) + o(n^{-1}).$$

In time series analysis we define a statistical curvature γ_θ which is a counterpart of Efron's statistical curvature defined in i.i.d. case. The relations between γ_θ and higher order asymptotic efficiency are illuminated. Also we investigate some asymptotic properties of an ancillary statistic which is constructed of $L_n^{(2)}(\mathbf{X}_n, \hat{\theta}_{ML}^{(n)})$. It is shown that the information lost by the reduction of $\mathbf{X}_n$ to $\hat{\theta}_{ML}^{(n)}$ is recovered by conditioning by the ancillary statistic.

We now proceed to discuss the above results in detail. Let $\mathbf{X}_n = (X_1, \ldots, X_n)'$ be a realization from a Gaussian stationary process having the spectral density $f_\theta(\lambda)$ depending on an unknown parameter $\theta \in \Theta \subset R^1$. Let $P_{n,\theta}$ denote the probability distribution of $\mathbf{X}_n$ on $(R^n, \mathbf{B}^n)$, where $\mathbf{B}^n$ is the Borel σ-algebra of R^n. We use the notations $\mathbf{D}_d$ and $\mathbf{D}_{ARMA}^c$, spaces of functions on $[-\pi, \pi]$ defined in Section 3.2. Here we require the following assumptions.

Assumption 4.1.1. $\{X_t\}$ is a Gaussian stationary process with the spectral density $f_\theta(\lambda) \in \mathbf{D}_{ARMA}^c$, $\theta \in \Theta \subset R^1$, and mean 0, where Θ is an open set of R^1.

Assumption 4.1.2. The spectral density $f_\theta(\lambda)$ is continuously five times differentiable with respect to $\theta \in \Theta$, and the derivatives $\partial f_\theta / \partial \theta, \ldots, \partial^5 f_\theta / \partial \theta^5$ belong to $\mathbf{D}_d$.

Assumption 4.1.3. For every compact set $C \subset \Theta$ there exists $d_1 > 0$ such that

$$I(\theta) = \frac{1}{4\pi} \int_{-\pi}^{\pi} \left\{ \frac{\partial}{\partial \theta} \log f_\theta(\lambda) \right\}^2 d\lambda \geq d_1 > 0,$$

for all $\theta \in C$.

The likelihood function based on $\mathbf{X}_n$ is given by

$$p_n(\mathbf{X}_n, \theta) = (2\pi)^{-\frac{n}{2}} |\Sigma_n|^{-\frac{1}{2}} \exp\left[-\frac{1}{2} \mathbf{X}_n' \Sigma_n^{-1} \mathbf{X}_n \right],$$

where Σ_n is the covariance matrix of $\mathbf{X}_n$. Define $L_n(\mathbf{X}_n, \theta) = \log p_n(\mathbf{X}_n, \theta)$. We estimate θ by the generalized maximum likelihood estimator $\hat{\theta}_{GML}^{(n)}$ defined in (3.2.2). First, we state

Lemma 4.1.1. *Suppose that Assumptions 4.1.1 – 4.1.3 hold. Then, for every integer $\gamma \geq 1$ and for every compact set $C \subset \Theta$, there exists $\epsilon > 0$ such that*

$$\sup_{\theta \in C} E_\theta \left\{ \sup_{|\tau - \theta| \leq \epsilon} \frac{1}{n} \left| \frac{\partial^5}{\partial \tau^5} L_n(\mathbf{X}_n, \tau) \right| \right\}^{2\gamma} < \infty.$$

Proof. Notice that

$$\frac{\partial^5}{\partial \tau^5} L_n(\mathbf{X}_n, \tau) = \mathbf{X}_n' A(\tau) \mathbf{X}_n + \text{tr} B(\tau), \tag{4.1.1}$$

where $A(\tau)$ and $B(\tau)$ are polynomials of $G_1(\tau) = \Sigma_n^{-1}$, $G_2(\tau) = \frac{\partial}{\partial \tau} \Sigma_n, \ldots, G_6(\tau) = \frac{\partial^5}{\partial \tau^5} \Sigma_n$. It follows that

$$\frac{1}{n} \left| \frac{\partial^5}{\partial \tau^5} L_n(\mathbf{X}_n, \tau) \right| \leq \frac{1}{n} \mathbf{X}_n' \mathbf{X}_n \cdot \|A(\tau)\| + \|B(\tau)\|, \tag{4.1.2}$$

where $\|A\|$ is the square root of the largest eigen value of AA'. Since $G_1(\tau), \ldots, G_6(\tau)$ are norm bounded on compact set (see the proof of Theorem 2.2.1), so are $A(\tau)$ and $B(\tau)$. Gaussianity of $\mathbf{X}_n$ and (4.1.2) imply our assertion.

Turning to the asymptotic sufficiency, we have

Theorem 4.1.1. *Suppose that Assumptions 4.1.1 – 4.1.3 and Assumptions 3.2.4 and 3.2.5 hold. Then the statistic*

$$t_n = \left(\hat{\theta}_{GML}^{(n)}, \frac{\partial L_n(\mathbf{X}_n, \hat{\theta}_{GML}^{(n)})}{\partial \theta}, \frac{\partial^2 L_n(\mathbf{X}_n, \hat{\theta}_{GML}^{(n)})}{\partial \theta^2}, \frac{\partial^3 L_n(\mathbf{X}_n, \hat{\theta}_{GML}^{(n)})}{\partial \theta^3}, \frac{\partial^4 L_n(\mathbf{X}_n, \hat{\theta}_{GML}^{(n)})}{\partial \theta^4} \right)$$

is asymptotically sufficient up to order $o(n^{-1})$ in the sense that
(i) t_n is sufficient for a family $\{Q_{n,\theta} : \theta \in \Theta\}$ of probability measures on $(R^n, \mathbf{B}^n)$,
(ii) for every compact set $C \subset \Theta$,

$$\sup_{\theta \in C} \|P_{n,\theta} - Q_{n,\theta}\| = o(n^{-1}). \tag{4.1.3}$$

Proof. Let $\mathbf{x}_n = (x_1, \ldots, x_n)'$ be the point in R^n which corresponds to the random vector $\mathbf{X}_n$. Using Taylor's formula we have

$$L_n(\mathbf{x}_n, \theta) = L_n(\mathbf{x}_n, \hat{\theta}_{GML}^{(n)}) + \sum_{l=1}^{4} \frac{L_n^{(l)}(\mathbf{x}_n, \hat{\theta}_{GML}^{(n)})(\theta - \hat{\theta}_{GML}^{(n)})^l}{l!} + R_n(\mathbf{x}_n, \theta), \tag{4.1.4}$$

where

$$R_n(\mathbf{x}_n, \theta) = n(\theta - \hat{\theta}_{GML}^{(n)})^5 \int_0^1 \frac{(1-v)^4}{4!} \frac{1}{n} L_n^{(5)} \left\{ \mathbf{x}_n, \hat{\theta}_{GML}^{(n)} + v(\theta - \hat{\theta}_{GML}^{(n)}) \right\} dv,$$

and

$$L_n^{(l)}(\mathbf{x}_n, \theta) = \frac{\partial^l}{\partial \theta^l} L_n(\mathbf{x}_n, \theta), \quad l = 1, \ldots, 5.$$

We set down

$$A_n = \left\{ \mathbf{x}_n \in R^n : \left| \theta - \hat{\theta}_{GML}^{(n)}(\mathbf{x}_n) \right| \le n^{\beta - \frac{1}{2}} \right\},$$
$$B_n = \left\{ \mathbf{x}_n \in R^n : b_n(\mathbf{x}_n) \le n^{\delta} \right\},$$

where

$$b_n(\mathbf{x}_n) = \sup \left\{ \left| \frac{1}{n} L_n^{(5)}(\mathbf{x}_n, \tau) \right| : \tau \in \Theta, \left| \tau - \hat{\theta}_{GML}^{(n)}(\mathbf{x}_n) \right| \le n^{\beta - \frac{1}{2}} \right\}.$$

Here β and δ are chosen so that $0 < \beta < \frac{1}{10}$, $0 < \delta < \frac{1}{2}$ and $5\beta + \delta < \frac{1}{2}$. In view of Theorem 3.2.1 we can see

$$\sup_{\theta \in C} P_{n,\theta}[A_n^c] = o(n^{-1}). \tag{4.1.5}$$

By Lemma 4.1.1 and Tchebychev's inequality there exists $\epsilon > 0$ such that

$$\sup_{\theta \in C} P_{n,\theta} \left[\sup_{|\tau - \theta| \le \epsilon} \frac{1}{n} \left| L_n^{(5)}(\mathbf{X}_n, \tau) \right| \ge n^{\delta} \right] = O(n^{-2\gamma\delta}).$$

Since we can choose the above γ so that $2\gamma\delta > 1$, we have

$$\sup_{\theta \in C} P_{n,\theta}[A_n \cap B_n^c] = o(n^{-1}). \tag{4.1.6}$$

Define

$$q_n(\mathbf{x}_n, \theta) = a_n(\theta)\chi_{A_n \cap B_n}(\mathbf{x}_n) \exp\left[L_n(\mathbf{x}_n, \hat{\theta}_{GML}^{(n)}) + H_n(t_n, \theta)\right],$$

where

$$H_n(t_n, \theta) = \sum_{l=1}^{4} \frac{L_n^{(l)}(\mathbf{x}_n, \hat{\theta}_{GML}^{(n)})(\theta - \hat{\theta}_{GML}^{(n)})^l}{l!},$$

$$a_n(\theta)^{-1} = \int_{R^n} \chi_{A_n \cap B_n}(\mathbf{x}_n) \exp\left[L_n(\mathbf{x}_n, \hat{\theta}_{GML}^{(n)}) + H_n(t_n, \theta)\right] d\mathbf{x}_n,$$

and χ_A is an indicator of the event A. Let $Q_{n,\theta}(A) = \int_A q_n(\mathbf{x}_n, \theta)\, d\mathbf{x}_n$, for $A \in \mathbf{B}^n$, then the statistic t_n is sufficient for $\{Q_{n,\theta}; \theta \in \Theta\}$ by the factorization theorem. Also, let

$$q_n^\star(\mathbf{x}_n, \theta) = \chi_{A_n \cap B_n}(\mathbf{x}_n) \exp\left[L_n(\mathbf{x}_n, \hat{\theta}_{GML}^{(n)}) + H_n(t_n, \theta)\right],$$

and

$$Q_{n,\theta}^\star(A) = \int_A q_n^\star(\mathbf{x}_n, \theta)\, d\mathbf{x}_n, \quad \text{for } A \in \mathbf{B}^n.$$

Then

$$
\begin{aligned}
\left\|P_{n,\theta} - Q_{n,\theta}^\star\right\| &= \frac{1}{2}\int_{R^n} |p_n(\mathbf{x}_n, \theta) - q_n^\star(\mathbf{x}_n, \theta)|\, d\mathbf{x}_n \\
&= \frac{1}{2}\int \chi_{A_n \cap B_n}(\mathbf{x}_n)\, |1 - \exp\{-R_n(\mathbf{x}_n, \theta)\}|\, p_n(\mathbf{x}_n, \theta)\, d\mathbf{x}_n \\
&\quad + \frac{1}{2}P_{n,\theta}\left[A_n^c\right] + \frac{1}{2}P_{n,\theta}\left[A_n \cap B_n^c\right].
\end{aligned}
\tag{4.1.7}
$$

For sufficiently large n, and for every $\mathbf{x}_n \in A_n \cap B_n$, it is easy to see that

$$|R_n(\mathbf{x}_n, \theta)| = O(n^{1+5(\beta-1/2)+\delta}) = o(n^{-1}),$$

uniformly in $\theta \in C$, which implies

$$
\begin{aligned}
\sup_{\theta \in C} \int \chi_{A_n \cap B_n}(\mathbf{x}_n)\, &|1 - \exp\{-R_n(\mathbf{x}_n, \theta)\}|\, p_n(\mathbf{x}_n, \theta)\, d\mathbf{x}_n \\
&\leq \sup_{\theta \in C} \sup_{\mathbf{x}_n \in A_n \cap B_n} |1 - \exp\{-R_n(\mathbf{x}_n, \theta)\}| \\
&\leq \sup_{\theta \in C} \sup_{\mathbf{x}_n \in A_n \cap B_n} |R_n(\mathbf{x}_n, \theta)| \exp\{|R_n(\mathbf{x}_n, \theta)|\} = o(n^{-1}).
\end{aligned}
\tag{4.1.8}
$$

It follows from (4.1.5), (4.1.6), (4.1.7) and (4.1.8) that

$$\sup_{\theta \in C} \left\|P_{n,\theta} - Q_{n,\theta}^\star\right\| = o(n^{-1}).$$

Finally we notice that

$$
\begin{aligned}
\sup_{\theta \in C} \left\|Q_{n,\theta}^\star - Q_{n,\theta}\right\| &= \sup_{\theta \in C} \frac{1}{2}\int_{R^n} |q_n^\star(\mathbf{x}_n, \theta) - q_n(\mathbf{x}_n, \theta)|\, d\mathbf{x}_n \\
&\leq \frac{1}{2}\sup_{\theta \in C} |a_n(\theta)^{-1} - 1| \\
&= \frac{1}{2}\sup_{\theta \in C} |a_n(\theta)^{-1}Q_{n,\theta}(R^n) - P_{n,\theta}(R^n)| \\
&= \frac{1}{2}\sup_{\theta \in C} |Q_{n,\theta}^\star(R^n) - P_{n,\theta}(R^n)| = o(n^{-1}),
\end{aligned}
$$

which completes the proof.

Corollary 4.1.1. Suppose that Assumptions 4.1.1 – 4.1.3 hold in the special case of $\hat{\theta}^{(n)}_{GML} = \hat{\theta}^{(n)}_{ML}$ (maximum likelihood estimator) which is a solution of $L^{(1)}_n(\mathbf{X}_n, \theta) = 0$, $\theta \in \Theta$. Then the statistic

$$t_n = \left(\hat{\theta}^{(n)}_{ML}, L^{(2)}_n\left(\mathbf{X}_n, \hat{\theta}^{(n)}_{ML}\right), L^{(3)}_n\left(\mathbf{X}_n, \hat{\theta}^{(n)}_{ML}\right), L^{(4)}_n\left(\mathbf{X}_n, \hat{\theta}^{(n)}_{ML}\right) \right)$$

is asymptotically sufficient up to order $o(n^{-1})$.

Now we state an application of Theorem 4.1.1. Let $W(\theta, \cdot)$ be a loss function such that for each $\theta \in \Theta$, $W(\theta, \cdot)$ is a measurable convex function, and for $C = [-a, a] \subset \Theta$, $a > 0$, $0 \leq W(\theta, y) \leq K < \infty$, for all $(\theta, y) \in C \times C$. Let U be a class of estimators $\hat{\theta}_n$ of θ satisfying the following conditions:

(i) $\hat{\theta}_n \in C$,

(ii) $E_\theta(\hat{\theta}_n) - \theta = o(n^{-1})$,   uniformly for $\theta \in C$,

(i.e., $\hat{\theta}_n$ is asymptotically unbiased up to order $o(n^{-1})$). Asymptotic modification of the Rao-Blackwell theorem yields (cf. Suzuki(1978)),

Theorem 4.1.2. *Suppose that Assumptions 4.1.1 – 4.1.3 and Assumptions 3.2.4 and 3.2.5 hold. If $\hat{\theta}_n \in U$, there exists an estimator $\hat{\Psi}(t_n)$ depending only on t_n such that $\hat{\Psi}(t_n) \in U$ and*

$$E_\theta W\left(\theta, \hat{\Psi}(t_n)\right) \leq EW\left(\theta, \hat{\theta}_n\right) + o(n^{-1}),$$

uniformly for $\theta \in C$.

We saw that the derivatives of log-likelihood at $\hat{\theta}^{(n)}_{GML}$ supply some additional informations which the estimator $\hat{\theta}^{(n)}_{GML}$ can not recover. Making an asymptotic ancillary statistic from $L^{(2)}_n(\mathbf{X}_n, \hat{\theta}^{(n)}_{ML})$ we shall discuss its asymptotic properties, and investigate a relation between higher order asymptotic efficiency and ancillary statistic. Let

$$
\begin{aligned}
Z_1(\theta) &= \frac{1}{\sqrt{n}} L^{(1)}_n(\mathbf{X}_n, \theta), \\
Z_2(\theta) &= \frac{1}{\sqrt{n}} \left\{ L^{(2)}_n(\mathbf{X}_n, \theta) - E_\theta L^{(2)}_n(\mathbf{X}_n, \theta) \right\}, \\
Z_3(\theta) &= \frac{1}{\sqrt{n}} \left\{ L^{(3)}_n(\mathbf{X}_n, \theta) - E_\theta L^{(3)}_n(\mathbf{X}_n, \theta) \right\}.
\end{aligned}
$$

We use the notations $J(\theta)$, $K(\theta)$, $L(\theta)$, $M(\theta)$, $N(\theta)$ and $H(\theta)$ defined in Lemma 2.3.1. In Section 2.3 we have seen that

$$\hat{\theta}^*_{ML} = \hat{\theta}^{(n)}_{ML} + \frac{K(\hat{\theta}^{(n)}_{ML})}{6nI(\hat{\theta}^{(n)}_{ML})^2},$$

is third-order AMU and satisfies the relation;

$$F_\theta^{(3)}(x) - P_{n,\theta}\left\{\sqrt{nI(\theta)}\left(\hat{\theta}_{ML}^* - \theta\right) \leq x\right\}$$

$$= \frac{x^3\phi(x)}{8nI(\theta)^3}\left\{M(\theta)I(\theta) - J(\theta)^2\right\} + o(n^{-1}), \quad \text{for } x > 0, \tag{4.1.9}$$

where $F_\theta^{(3)}$ is the third-order bound distribution and $\phi(x) = (1/\sqrt{2\pi})\exp[-x^2/2]$ (see (2.3.45)). We put

$$\gamma_\theta = \frac{\{M(\theta)I(\theta) - J(\theta)^2\}^{1/2}}{I(\theta)^{3/2}}.$$

For i.i.d. case Efron(1975) introduced the statistical curvature in a differential-geometrical framework. Our γ_θ is a counterpart of the statistical curvature in time series analysis. The relation (4.1.9) implies that our γ_θ is a measure of a certain amount of information which the maximum likelihood estimator can not recover. Since $E_\theta(Z_2(\theta)^2) = M(\theta) + O(n^{-1})$, $\text{Cov}_\theta(Z_1(\theta), Z_2(\theta)) = J(\theta) + O(n^{-1})$ and $E_\theta(Z_1(\theta)^2) = I(\theta) + O(n^{-1})$ (see Lemma 2.3.1) we have

$$E_\theta\left\{Z_2(\theta) - \frac{J(\theta)}{I(\theta)}Z_1(\theta)\right\}^2 = M(\theta) - \frac{J(\theta)^2}{I(\theta)} + O(n^{-1})$$

$$= \{I(\theta)\gamma_\theta\}^2 + O(n^{-1}). \tag{4.1.10}$$

The relations (4.1.9) and (4.1.10) imply that the following three statements are equivalent:

(i) $\hat{\theta}_{ML}^*$ is third-order asymptotically efficient among third-order AMU estimators,

(ii) $\gamma_\theta = 0$,

(iii) $Z_2(\theta) - \frac{J(\theta)}{I(\theta)}Z_1(\theta) \to 0$, in quadratic mean.

For i.i.d case Efron and Hinkley(1978) investigated various asymptotic properties of the observed Fisher information. Skovgaard(1985) gave an asymptotic expansion of the distribution of the maximum likelihood estimator given an ancillary statistic, which is defined as a standardized version of the observed Fisher information. We can extend this type of approach to our dependent case. Put

$$\hat{B} = \sqrt{n}\left\{-\frac{1}{n}L_n^{(2)}\left(\mathbf{X}_n, \hat{\theta}_{ML}^{(n)}\right) - I\left(\hat{\theta}_{ML}^{(n)}\right)\right\}.$$

Expanding the right hand side of $\hat{B}$ around $\hat{\theta}_{ML}^{(n)} = \theta$ we get

$$\hat{B} = -\sqrt{n}\left\{\frac{1}{n}L_n^{(2)}(\mathbf{X}_n, \theta) - E_\theta\frac{1}{n}L_n^{(2)}(\mathbf{X}_n, \theta)\right\}$$

$$- \sqrt{n}\left\{E_\theta\frac{1}{n}L_n^{(2)}(\mathbf{X}_n, \theta) + I(\theta)\right\} - \frac{1}{\sqrt{n}}Z_3(\theta)\sqrt{n}\left(\hat{\theta}_{ML}^{(n)} - \theta\right)$$

$$- \sqrt{n}\left(\hat{\theta}_{ML}^{(n)} - \theta\right)\left\{\frac{\partial}{\partial\theta}I(\theta) + E_\theta\frac{1}{n}L_n^{(3)}(\mathbf{X}_n, \theta)\right\}$$

$$- \frac{1}{2\sqrt{n}}\left\{\sqrt{n}\left(\hat{\theta}_{ML}^{(n)} - \theta\right)\right\}^2\left\{E_\theta\frac{1}{n}L_n^{(4)}(\mathbf{X}_n, \theta) + \frac{\partial^2}{\partial\theta^2}I(\theta)\right\} + o_p(n^{-\frac{1}{2}}). \tag{4.1.11}$$

Remember that

$$\sqrt{n}\left(\hat{\theta}_{ML}^{(n)} - \theta\right) = \frac{Z_1}{I} + \frac{1}{I^2\sqrt{n}}\left\{Z_1 Z_2 - \frac{3J + K}{2I}Z_1^2\right\} + o_p(n^{-\frac{1}{2}}),$$

$$\frac{\partial}{\partial\theta}I(\theta) = 2J(\theta) + K(\theta), \quad \frac{\partial}{\partial\theta}J(\theta) = L(\theta) + M(\theta) + N(\theta),$$

$$\frac{\partial}{\partial\theta}K(\theta) = 3N(\theta) + H(\theta), \quad E_\theta \frac{1}{n}L_n^{(3)}(\mathbf{X}_n, \theta) = -3J(\theta) - K(\theta) + O(n^{-1}),$$

and

$$E_\theta \frac{1}{n}L_n^{(4)}(\mathbf{X}_n, \theta) = -4L(\theta) - 3M(\theta) - 6N(\theta) - H(\theta) + O(n^{-1}),$$

(see Section 2.3). Then we have

$$\hat{B} = -\left(Z_2 - \frac{J}{I}Z_1\right) + \frac{\Delta}{\sqrt{n}} - \frac{Z_1}{I\sqrt{n}}\left\{Z_3 - \frac{J}{I}Z_2\right\}$$

$$- \frac{Z_1^2}{2I^2\sqrt{n}}\left\{-2L - M - N + \frac{J(3J + K)}{I}\right\} + o_p(n^{-\frac{1}{2}}), \qquad (4.1.12)$$

where Δ is defined by $E_\theta Z_1^2 = I + \frac{\Delta}{n} + o(n^{-1})$. From (4.1.10), the above (4.1.12) implies that $\hat{B}$ is a first-order approximation to the residual in $Z_2(\theta)$ after linear regression on $Z_1(\theta)$, whose variance is $\gamma_\theta^2 I(\theta)^2$. Thus we have

Theorem 4.1.3. *Under Assumptions 4.1.1 – 4.1.3, the distribution of the statistic $\hat{A} = \{I(\hat{\theta}_{ML}^{(n)})\gamma_{\hat{\theta}_{ML}^{(n)}}\}^{-1}\hat{B}$, as $n \to \infty$, tends to the normal distribution $N(0, 1)$.*

Theorem 4.1.3 implies that $\hat{A}$ is a first-order asymptotic ancillary statistic. In the next section we will derive the Edgeworth expansion of $\hat{A}$ up to second order.

Now we show that the information lost by the reduction of $\mathbf{X}_n$ to $\hat{\theta}_{ML}^{(n)}$ is recovered by conditioning by the ancillary statistic $\hat{A}$. In time series situation Hosoya(1979) gave the relation;

$$\lim_{n\to\infty} E\left[\text{Var}\left\{\frac{\partial}{\partial\theta}L_n(\mathbf{X}_n, \theta)\bigg|\sqrt{n}\left(\hat{\theta}_{ML}^{(n)} - \theta\right)\right\}\right] = \gamma_\theta^2 I(\theta), \qquad (4.1.13)$$

which means that the maximum likelihood estimator of a spectral parameter is second-order asymptotically efficient in Rao's sense. The asymptotic ancillary statistic $\hat{A}$ recovers further information in the following sense.

Theorem 4.1.4. *Under Assumptions 4.1.1 – 4.1.3, we have*

$$E\left[\text{Var}\left\{\frac{\partial}{\partial\theta}L_n(\mathbf{X}_n, \theta)\bigg|\sqrt{n}\left(\hat{\theta}_{ML}^{(n)} - \theta\right), \hat{A}\right\}\right] = O(n^{-1}). \qquad (4.1.14)$$

Proof. Expanding $\frac{\partial}{\partial\theta}L_n(\mathbf{X}_n, \hat{\theta}_{ML}^{(n)}) = 0$, at $\hat{\theta}_{ML}^{(n)} = \theta$, we have

$$\frac{\partial}{\partial\theta}L_n(\mathbf{X}_n, \theta) = -\left(\hat{\theta}_{ML}^{(n)} - \theta\right)L_n^{(2)}(\mathbf{X}_n, \theta) - \frac{1}{2}\left(\hat{\theta}_{ML}^{(n)} - \theta\right)^2 L_n^{(3)}(\mathbf{X}_n, \theta) + O_p(n^{-\frac{1}{2}})$$

$$
\begin{aligned}
= & -\left(\hat{\theta}_{ML}^{(n)} - \theta\right)\left\{L_n^{(2)}\left(\mathbf{X}_n, \hat{\theta}_{ML}^{(n)}\right) + \left(\theta - \hat{\theta}_{ML}^{(n)}\right)L_n^{(3)}\left(\mathbf{X}_n, \hat{\theta}_{ML}^{(n)}\right)\right\} \\
& -\frac{1}{2}\left(\hat{\theta}_{ML}^{(n)} - \theta\right)^2 E_\theta L_n^{(3)}(\mathbf{X}_n, \theta) + O_p(n^{-\frac{1}{2}}) \\
= & -\left(\hat{\theta}_{ML}^{(n)} - \theta\right)L_n^{(2)}\left(\mathbf{X}_n, \hat{\theta}_{ML}^{(n)}\right) \\
& +\frac{1}{2}\left\{\sqrt{n}\left(\hat{\theta}_{ML}^{(n)} - \theta\right)\right\}^2\{-3J(\theta) - K(\theta)\} + O_p(n^{-\frac{1}{2}}).
\end{aligned}
\tag{4.1.15}
$$

From the definition of $\hat{A}$ we have

$$
-\sqrt{n}\left\{I\left(\hat{\theta}_{ML}^{(n)}\right)\gamma_{\hat{\theta}_{ML}^{(n)}}\right\}\hat{A} - nI\left(\hat{\theta}_{ML}^{(n)}\right) = L_n^{(2)}\left(\mathbf{X}_n, \hat{\theta}_{ML}^{(n)}\right).
\tag{4.1.16}
$$

Substituting (4.1.16) for (4.1.15) we have

$$
\begin{aligned}
\frac{\partial}{\partial\theta}L_n(\mathbf{X}_n, \theta) = & \sqrt{n}\left(\hat{\theta}_{ML}^{(n)} - \theta\right)\left[\left\{I\left(\hat{\theta}_{ML}^{(n)}\right)\gamma_{\hat{\theta}_{ML}^{(n)}}\right\}\hat{A} + \sqrt{n}I\left(\hat{\theta}_{ML}^{(n)}\right)\right] \\
& +\frac{1}{2}\left\{\sqrt{n}\left(\hat{\theta}_{ML}^{(n)} - \theta\right)\right\}^2\{-3J(\theta) - K(\theta)\} + O_p(n^{-\frac{1}{2}}) \\
= & \sqrt{n}\left(\hat{\theta}_{ML}^{(n)} - \theta\right)\left[\left\{I(\theta)\gamma_\theta + \left(\hat{\theta}_{ML}^{(n)} - \theta\right)\frac{\partial}{\partial\theta}(I(\theta)\gamma_\theta)\right\}\hat{A} + \sqrt{n}I(\theta)\right. \\
& \left. +\sqrt{n}\left(\hat{\theta}_{ML}^{(n)} - \theta\right)\frac{\partial}{\partial\theta}I(\theta) + \frac{1}{2\sqrt{n}}n\left(\hat{\theta}_{ML}^{(n)} - \theta\right)^2\frac{\partial^2}{\partial\theta^2}I(\theta)\right] \\
& +\frac{1}{2}\left\{\sqrt{n}\left(\hat{\theta}_{ML}^{(n)} - \theta\right)\right\}^2\{-3J(\theta) - K(\theta)\} + O_p(n^{-\frac{1}{2}}),
\end{aligned}
$$

which shows that the conditional variance of $\partial L_n(\mathbf{X}_n, \theta)/\partial\theta$ given $\{\sqrt{n}(\hat{\theta}_{ML}^{(n)} - \theta), \hat{A}\}$ is $O(n^{-1})$.

4.2. Asymptotic ancillarity in time series analysis

As we saw in the previous section, an asymptotic ancillary statistic is constructed in terms of the maximum likelihood estimator and the observed Fisher information. Here we shall show that this ancillary statistic is second-order locally ancillary. Then we give a sufficient condition that the modified ancillary statistic is second-order asymptotically ancillary. Also the second-order Edgeworth expansion of the conditional distribution of the maximum likelihood estimator given the ancillary statistic is evaluated. Using this Edgeworth expansion we can calculate confidence intervals with probability levels to $O(n^{-1})$, where n is the sample size. Some numerical studies are made and they show that our second-order approximation for the distribution of the ancillary statistic is good.

Let $\{X_t\}$ be a Gaussian ARMA process with the spectral density $f_\theta(\lambda)$, where θ is an unknown parameter. Throughout this section we use the same notations as those of Section 4.1, and assume the Assumptions 4.1.1 – 4.1.3.

For i.i.d. case Efron and Hinkley(1978) investigated various asymptotic properties of the observed Fisher information. Cox(1980) introduced the concept of local ancillarity. Skovgaard(1985) gave an asymptotic expansion of the distribution of the maximum likelihood estimator given an ancillary statistic, which is defined as a standardized version of the observed Fisher information.

Here we shall extend this type of approach to our Gaussian ARMA processes. Remember the ancillary statistic

$$\hat{A} = \left\{ I\left(\hat{\theta}_{ML}^{(n)}\right) \gamma_{\hat{\theta}_{ML}^{(n)}} \right\}^{-1} \cdot \hat{B},$$

which is a standardized version of the observed Fisher information in time series analysis (see Theorem 4.1.3). In view of Theorem 4.1.1 and 4.1.4, $\hat{A}$ has a higher order information which the maximum likelihood estimator can not recover. Note that

$$\left\{ I\left(\hat{\theta}_{ML}^{(n)}\right) \gamma_{\hat{\theta}_{ML}^{(n)}} \right\}^{-1} = \{I(\theta)\gamma_\theta\}^{-1} - \frac{Z_1}{\sqrt{n}I(\theta)^3\gamma_\theta^2} \frac{\partial}{\partial\theta}\{I(\theta)\gamma_\theta\} + o_p(n^{-\frac{1}{2}}). \tag{4.2.1}$$

It follows from (4.1.12) and (4.2.1) that

$$\begin{aligned}
\hat{A} = {}& -\frac{1}{I\gamma_\theta}\left(Z_2 - \frac{J}{I}Z_1\right) + \frac{\Delta}{I\gamma_\theta\sqrt{n}} - \frac{Z_1}{I^2\gamma_\theta\sqrt{n}}\left(Z_3 - \frac{J}{I}Z_2\right) \\
& -\frac{Z_1^2}{2I^3\gamma_\theta\sqrt{n}}\left\{-2L - M - N + \frac{J(3J+K)}{I}\right\} \\
& +\frac{1}{I^3\gamma_\theta^2\sqrt{n}}\frac{\partial}{\partial\theta}\{I\gamma_\theta\}Z_1\left(Z_2 - \frac{J}{I}Z_1\right) + o_p(n^{-\frac{1}{2}}).
\end{aligned} \tag{4.2.2}$$

Since the asymptotic moments for Z_1, Z_2 and Z_3 are evaluated in Lemma 2.3.1, it is not difficult to show that

$$E_\theta(\hat{A}) = \frac{1}{\sqrt{n}}b(\theta) + O(n^{-1}), \tag{4.2.3}$$

$$\mathrm{cum}_\theta\{\hat{A}, \hat{A}\} = 1 + O(n^{-1}), \tag{4.2.4}$$

$$\mathrm{cum}_\theta\{\hat{A}, \hat{A}, \hat{A}\} = \frac{1}{\sqrt{n}}\kappa_3(\theta) + O(n^{-1}), \tag{4.2.5}$$

$$\mathrm{cum}_\theta^{(J)}\{\hat{A}, \ldots, \hat{A}\} = O(n^{-\frac{J}{2}+1}), \quad \text{for } J \geq 3. \tag{4.2.6}$$

where

$$b(\theta) = \frac{1}{2I(\theta)^3\gamma_\theta}\left[2I(\theta)^2\Delta(\theta) + I(\theta)\left\{M(\theta) + N(\theta)\right\} - J(\theta)^2 - J(\theta)K(\theta)\right],$$

and

$$\kappa_3(\theta) = \frac{1}{2\pi\{I(\theta)\gamma_\theta\}^3}\int_{-\pi}^{\pi}\left[\left\{\frac{\partial^2}{\partial\theta^2}f_\theta(\lambda)^{-1}\right\}f_\theta(\lambda) - \frac{J(\theta)}{I(\theta)}\left\{\frac{\partial}{\partial\theta}f_\theta(\lambda)^{-1}\right\}f_\theta(\lambda)\right]^3 d\lambda.$$

From (2.1.16) we get,

Theorem 4.2.1. *The Edgeworth expansion for $\hat{A}$ is given by*

$$P_\theta[\hat{A} < y] = \int_{-\infty}^{y}\phi(x)\left[1 + \frac{1}{\sqrt{n}}b(\theta)x + \frac{1}{6\sqrt{n}}\kappa_3(\theta)(x^3 - 3x)\right]dx + O(n^{-1}). \tag{4.2.7}$$

Furthermore if $b(\theta)$ and $\kappa_3(\theta)$ are continuously differentiable, then, for $\delta > 0$,

$$P_{\theta_0+\delta/\sqrt{n}}[\hat{A} < y] = \int_{-\infty}^{y}\phi(x)\left[1 + \frac{1}{\sqrt{n}}b(\theta_0)x + \frac{1}{6\sqrt{n}}\kappa_3(\theta_0)(x^3 - 3x)\right]dx + O_\delta(n^{-1}), \tag{4.2.8}$$

which implies that $\hat{A}$ is second-order locally ancillary at $\theta = \theta_0$ in the sense of Cox(1980).

Now we consider to make a second-order asymptotic ancillary statistic. We modify $\hat{A}$ so that

$$\hat{A}^* = \hat{A} - \frac{1}{\sqrt{n}}b(\hat{\theta}_{ML}^{(n)}) = \hat{A} - \frac{1}{\sqrt{n}}b(\theta) + O_p(n^{-1}).$$

Similarly we have

Theorem 4.2.2. *The modified ancillary statistic $\hat{A}^*$ has the Edgeworth expansion*

$$P_\theta[\hat{A}^* < y] = \int_{-\infty}^{y} \phi(x)\left[1 + \frac{1}{6\sqrt{n}}\kappa_3(\theta)(x^3 - 3x)\right] dx + O(n^{-1}). \tag{4.2.9}$$

Thus if $\kappa_3(\theta)$ is independent of θ, our $\hat{A}^*$ is a second-order asymptotically ancillary statistic.

Example 4.2.1. Let $\{X_t\}$ be an autoregressive process of order 1, which is defined as $X_t = \alpha X_{t-1} + \epsilon_t$, $\epsilon_t \sim$ i.i.d. $N(0, \sigma^2)$ and $|\alpha| < 1$. For $\theta = \alpha$ we can see that $b(\alpha) = \sqrt{2}(2\alpha^2 - 1)/(1 - \alpha^2)$ and $\kappa_3(\alpha) = 2\sqrt{2}$. Theorem 4.2.2 implies that the modified ancillary statistic $\hat{A}^* = \hat{A} - \frac{1}{\sqrt{n}}b(\hat{\alpha}_{ML}^{(n)})$ is second-order asymptotically ancillary, and has the Edgeworth expansion;

$$P_\alpha[\hat{A}^* < y] = \int_{-\infty}^{y} \phi(x)\left[1 + \frac{\sqrt{2}}{3\sqrt{n}}(x^3 - 3x)\right] dx + O(n^{-1}). \tag{4.2.10}$$

However, if $\{X_t\}$ is a moving average process which is defined as $X_t = \epsilon_t - \beta\epsilon_{t-1}$, $|\beta| < 1$, then we can show that $\kappa_3(\beta)$ depends on β. In this case the modified ancillary statistic is not second-order asymptotically ancillary.

We next turn to give the Edgeworth expansions of the conditional distributions of $\{nI(\theta)\}^{-1/2}$ $L_n^{(1)}(\mathbf{X}_n, \theta)$ and $\sqrt{nI}(\hat{\theta}_{ML}^{(n)} - \theta)$ given $\hat{A}$. To avoid unnecessarily troublesome calculation, henceforth, we confine ourselves to the case when $\{X_t\}$ is generated by $X_t = \alpha X_{t-1} + \epsilon_t$, $|\alpha| < 1$, where the ϵ_ts are i.i.d. normal random variables with mean zero and variance σ^2. From (4.2.2) together with Proposition 2.3.1, the ancillary statistic $\hat{A}$ has the stochastic expansion;

$$\hat{A} = A_1 + \frac{1}{\sqrt{n}}Z_1 A_2 + \frac{1}{\sqrt{n}}C_1 + o_p(n^{-\frac{1}{2}}), \tag{4.2.11}$$

where

$$A_1 = -\frac{1-\alpha^2}{\sqrt{2}}\left\{Z_2 + \frac{2\alpha}{1-\alpha^2}Z_1\right\},$$

$$A_2 = -\frac{(1-\alpha^2)^2}{\sqrt{2}}Z_3 - \frac{1-\alpha^2}{\sqrt{2}}Z_1,$$

and

$$C_1 = \frac{3\alpha^2 - 1}{\sqrt{2}(1-\alpha^2)}.$$

Putting $U_1 = \{I(\alpha)\}^{-1/2}Z_1(\alpha)$ and $U_2 = \hat{A}$, we can evaluate the joint cumulants of U_1 and U_2;

$$\text{cum}(U_1) = 0, \quad \text{cum}(U_2) = \frac{\sqrt{2}(2\alpha^2 - 1)}{\sqrt{n}(1 - \alpha^2)} + O(n^{-1}),$$

$$\text{cum}(U_1, U_2) = O(n^{-1}), \quad \text{cum}(U_1, U_1) = \text{cum}(U_2, U_2) = 1 + O(n^{-1}),$$

$$\text{cum}(U_1, U_1, U_1) = \frac{6\alpha}{\{n(1 - \alpha^2)\}^{1/2}} + O(n^{-1}), \quad \text{cum}(U_2, U_2, U_2) = \frac{2\sqrt{2}}{\sqrt{n}} + O(n^{-1}),$$

$$\text{cum}(U_1, U_1, U_2) = \frac{\sqrt{2}}{\sqrt{n}} + O(n^{-1}), \quad \text{cum}(U_1, U_2, U_2) = O(n^{-1}),$$

$$\text{cum}^{(J)}(U_{i_1}, \ldots, U_{i_J}) = O(n^{-\frac{J}{2}+1}), \quad \text{for } i_j = 1 \text{ or } 2, \text{ and for } J \geq 3.$$

Then we have

$$
\begin{aligned}
P_\alpha\{U_1 < u, \ \hat{A} < a\} &= \int_{-\infty}^u \int_{-\infty}^a \phi(y_1)\phi(y_2) \Bigg[1 + \frac{\sqrt{2}(2\alpha^2 - 1)}{\sqrt{n}(1 - \alpha^2)} y_2 \\
&\quad + \frac{1}{6\sqrt{n}} \bigg\{ \frac{6\alpha}{(1 - \alpha^2)^{1/2}}(y_1^3 - 3y_1) + 2\sqrt{2}(y_2^3 - 3y_2) \\
&\quad + 3\sqrt{2}(y_1^2 y_2 - y_2) \bigg\} \Bigg] \, dy_1 dy_2 + O(n^{-1}).
\end{aligned}
\tag{4.2.12}
$$

From (4.2.7) and (4.2.12) we get

Theorem 4.2.3. *The Edgeworth expansion of the conditional distribution function of U_1 given $\hat{A} = a$ is*

$$F(u|a) = \Phi(u) - \phi(u) \left[\frac{\alpha}{\sqrt{n}(1 - \alpha^2)}(u^2 - 1) + \frac{au}{\sqrt{2n}} \right] + O(n^{-1}). \tag{4.2.13}$$

For unconditional case the Edgeworth expansion of the distribution function of U_1 is

$$F(u) = \Phi(u) - \phi(u) \left[\frac{\alpha}{\sqrt{n}(1 - \alpha^2)}(u^2 - 1) \right] + O(n^{-1}). \tag{4.2.14}$$

Similarly, evaluating the asymptotic cumulants it is not difficult to show that the Edgeworth expansion of the joint density of $\sqrt{nI(\alpha)}(\hat{\alpha}_{ML}^{(n)} - \alpha)$ and $\hat{A}$ is given by

$$
\begin{aligned}
f(y_1, y_2) &= \phi(y_1)\phi(y_2) \Bigg[1 + \frac{1}{\sqrt{n}} \bigg\{ -\frac{2\alpha}{\sqrt{1 - \alpha^2}} y_1 + \frac{\sqrt{2}(2\alpha^2 - 1)}{1 - \alpha^2} y_2 \bigg\} \\
&\quad + \frac{1}{6\sqrt{n}} \bigg\{ -\frac{6\alpha}{\sqrt{1 - \alpha^2}}(y_1^3 - 3y_1) + 2\sqrt{2}(y_2^3 - 3y_2) \\
&\quad - 3\sqrt{2}\,(-y_2 + y_1^2 y_2) \bigg\} \Bigg] + O(n^{-1}),
\end{aligned}
\tag{4.2.15}
$$

which implies

Theorem 4.2.4. *The Edgeworth expansion of the conditional distribution function of* $\sqrt{nI(\alpha)}$ $(\hat{\alpha}_{ML}^{(n)} - \alpha)$ *given* $\hat{A} = a$ *is*

$$G(x|a) = \Phi(x) + \frac{1}{\sqrt{n}}\phi(x)\left\{\frac{2\alpha}{\sqrt{1-\alpha^2}} + \frac{\alpha}{\sqrt{1-\alpha^2}}(x^2-1) + \frac{ax}{\sqrt{2}}\right\} + O(n^{-1}). \qquad (4.2.16)$$

For unconditional case the Edgeworth expansion of the distribution function of $\sqrt{nI(\alpha)}(\hat{\alpha}_{ML}^{(n)} - \alpha)$ *is*

$$G(x) = \Phi(x) + \frac{1}{\sqrt{n}}\phi(x)\left\{\frac{2\alpha}{\sqrt{1-\alpha^2}} + \frac{\alpha}{\sqrt{1-\alpha^2}}(x^2-1)\right\} + O(n^{-1}). \qquad (4.2.17)$$

Using the Edgeworth expansions given in Theorems 4.2.3 and 4.2.4 we can calculate confidence intervals with probability levels to $O(n^{-1})$. Applying the Cornish-Fisher inversion to (4.2.13), we get the following $(1-\epsilon)$ confidence interval for U_1;

$$\{I(\alpha)\}^{-\frac{1}{2}} Z_1(\alpha) \in \left[-z_\epsilon\left(1 + \frac{a}{\sqrt{2n}}\right) + \frac{\alpha}{\sqrt{n(1-\alpha^2)}}(z_\epsilon^2 - 1),\right.$$

$$\left.z_\epsilon\left(1 + \frac{a}{\sqrt{2n}}\right) + \frac{\alpha}{\sqrt{n(1-\alpha^2)}}(z_\epsilon^2 - 1)\right], \qquad (4.2.18)$$

where $\Phi(-z_\epsilon)$. In our model $X_t = \alpha X_{t-1} + \epsilon_t$, without loss of generality, we assume the innovation variance σ^2 is equal to 1. Then,

$$\{I(\alpha)\}^{-\frac{1}{2}} Z_1(\alpha) = \left\{\frac{1-\alpha^2}{n}\right\}^{\frac{1}{2}} \left\{-\frac{\alpha}{1-\alpha^2} + P_1 - \alpha P_0\right\}, \qquad (4.2.19)$$

where $P_1 = \sum_{t=2}^n X_t X_{t-1}$ and $P_0 = \sum_{t=2}^{n-1} X_t^2$ (e.g., Anderson(1971, p.354)). By (4.2.19), the relation (4.2.18) is equivalent to

$$\left\{\frac{1-\alpha^2}{n}\right\}^{\frac{1}{2}} \{P_1 - \alpha P_0\}$$

$$\in \left[-z_\epsilon\left(1 + \frac{a}{\sqrt{2n}}\right) + \frac{\alpha z_\epsilon^2}{\sqrt{n(1-\alpha^2)}}, \ z_\epsilon\left(1 + \frac{a}{\sqrt{2n}}\right) + \frac{\alpha z_\epsilon^2}{\sqrt{n(1-\alpha^2)}}\right]. \qquad (4.2.20)$$

solving (4.2.20) with respect to α we have

Theorem 4.2.5. *From the Edgeworth expansion (4.2.13), the* $(1-\epsilon)$ *confidence interval for* α *with probability level to* $O(n^{-1})$ *is given by*

$$\alpha \in \left[\hat{\alpha} - \frac{z_\epsilon}{\hat{v}\{n(1-\hat{\alpha}^2)\}^{1/2}}\left(1 + \frac{a}{\sqrt{2n}}\right) + \frac{z_\epsilon^2\hat{\alpha}}{n\hat{v}^2(1-\hat{\alpha}^2)^2}(1 - \hat{v} + \hat{v}\hat{\alpha}^2),\right.$$

$$\left.\hat{\alpha} + \frac{z_\epsilon}{\hat{v}\{n(1-\hat{\alpha}^2)\}^{1/2}}\left(1 + \frac{a}{\sqrt{2n}}\right) + \frac{z_\epsilon^2\hat{\alpha}}{n\hat{v}^2(1-\hat{\alpha}^2)^2}(1 - \hat{v} + \hat{v}\hat{\alpha}^2)\right], \qquad (4.2.21)$$

where $\hat{\alpha} = P_1/P_0$ *and* $\hat{v} = P_0/n$. *In the unconditional distribution (4.2.14), the* $(1-\epsilon)$ *confidence interval for* α *with probability level to* $O(n^{-1})$ *is equal to the above interval (4.2.21) with* $a = 0$.

Thus if $a < 0$, the confidence interval (4.2.21) is shorter than that made from the unconditional distribution.

Similarly, applying the Cornish-Fisher inversions to (4.2.16) and (4.2.17) we have

Theorem 4.2.6. *From the Edgeworth expansion (4.2.16), the $(1 - \epsilon)$ confidence interval for α with probability level to $O(n^{-1})$ is given by*

$$
\alpha \in \left[\hat{\alpha}_{ML}^{(n)} - z_\epsilon \left\{ \frac{1 - \hat{\alpha}_{ML}^{(n)\,2}}{n} \right\}^{\frac{1}{2}} \left(1 - \frac{a}{\sqrt{2n}} \right) + \frac{\hat{\alpha}_{ML}^{(n)}}{n}, \right.
$$

$$
\left. \hat{\alpha}_{ML}^{(n)} + z_\epsilon \left\{ \frac{1 - \hat{\alpha}_{ML}^{(n)\,2}}{n} \right\}^{\frac{1}{2}} \left(1 - \frac{a}{\sqrt{2n}} \right) + \frac{\hat{\alpha}_{ML}^{(n)}}{n} \right].
$$

$$(4.2.22)$$

In the unconditional distribution (4.2.17), the $(1 - \epsilon)$ confidence interval for α with probability level to $O(n^{-1})$ is equal to the above interval (4.2.22) with $a = 0$. Thus if $a > 0$, the confidence interval (4.2.22) is shorter than that made from the unconditional distribution.

Remark 4.2.1. In practice, if $a < 0$ ($a \geq 0$) we use the conditional confidence interval (4.2.21) (the unconditional one) in Theorem 4.2.5. In Theorem 4.2.6, if $a > 0$ ($a \leq 0$) we use the conditional confidence interval (4.2.22) (the unconditional one).

Remark 4.2.2. From (4.2.3) we have $E_\theta(\hat{A}) = \sqrt{2}(2\alpha^2 - 1)/\sqrt{n}(1 - \alpha^2) + O(n^{-1})$. If $|\alpha| < 1/\sqrt{2}$, we can expect that the probability of the event $\hat{A} < 0$ is asymptotically greater than that of $\hat{A} \geq 0$. Thus if we get a partial prior information of α (e.g., the process concerned is near an ARI-process) this information will be useful to choose the conditional or unconditional confidence interval.

We next give some numerical investigations related to the approximation (4.2.10) in the autoregressive process $X_t = \alpha X_{t-1} + \epsilon_t$, where ϵ_t are i.i.d. $N(0, 1)$. Let

$$
\tilde{\alpha}_{ML}^{(n)} = (1 - n^{-1}) \frac{\sum_{t=2}^n X_t X_{t-1}}{\sum_{t=2}^{n-1} X_t^2}.
$$

It is known (Fujikoshi and Ochi(1984)) that

$$
P_\alpha \left\{ \sqrt{nI(\alpha)} \left(\hat{\alpha}_{ML}^{(n)} - \alpha \right) \leq x \right\} - P_\alpha \left\{ \sqrt{nI(\alpha)} \left(\tilde{\alpha}_{ML}^{(n)} - \alpha \right) \leq x \right\} = o(n^{-1}).
$$

Thus, henceforth, we can use $\tilde{\alpha}_{ML}^{(n)}$ in place of the exact maximum likelihood estimator $\hat{\alpha}_{ML}^{(n)}$ without disturbing the asymptotic theory. In (4.2.10) we can put

$$
\hat{A}^* = \frac{1}{\sqrt{2\pi}} \frac{1 + \tilde{\alpha}_{ML}^{(n)\,2}}{1 - \tilde{\alpha}_{ML}^{(n)\,2}} + \frac{1 - \tilde{\alpha}_{ML}^{(n)\,2}}{\sqrt{2n}} \sum_{t=2}^{n-1} X_t^2 - \sqrt{\frac{n}{2}} - \frac{\sqrt{2}}{\sqrt{n}} \frac{2\tilde{\alpha}_{ML}^{(n)\,2} - 1}{1 - \tilde{\alpha}_{ML}^{(n)\,2}}.
$$

Let

$$
F_\alpha(x) = P_\alpha[\hat{A}^* < x], \quad Nor(x) = \int_{-\infty}^x \phi(t)\,dt,
$$

and

$$Edg(x) = \int_{-\infty}^{x} \phi(t) \left[1 + \frac{\sqrt{2}}{3\sqrt{n}} (t^3 - 3t) \right] dt.$$

For $n = 100$ we computed $F_\alpha(\cdot)$ by 1000 trials simulation. We plotted $F_\alpha(x)$, $\alpha = 0.0, 0.3, 0.6, 0.9$, together with the graphs of $Nor(x)$ and $Edg(x)$ in Figures 4.2.1 – 4.2.4 in the Appendix. The results show that $Nor(x)$ and $Edg(x)$ approximate $F_\alpha(x)$ well, and that $Edg(x)$ gives a better approximation than $Nor(x)$ generally (although both approximations are not so good for the case $\alpha = 0.9$ near an explosive process).

We next examined Theorems 4.2.5 and 4.2.6 by simulation. In Table 4.2.1 of Appendix, I_1 and I_2 are the 0.95-confidence intervals (4.2.21) and (4.2.21) with $a = 0$, respectively, for $\alpha = -0.9$ (0.1) 0.9. Similarly, in Table 4.2.2, I_3 and I_4 are the 0.95-confidence intervals (4.2.22) and (4.2.22) with $a = 0$, respectively, for $\alpha = -0.9$ (0.1) 0.9 (all end points of I_1, I_2, I_3 and I_4 are the averaged values of five trials simulation). As we said in Remark 4.2.2, the conditional confidence interval I_1 is better than the unconditional one I_2 if $|\alpha|$ is small in comparison with 1. Also Table 4.2.2 shows that I_3 is better than I_4 if $|\alpha|$ is near 1, which confirms Remark 4.2.2.

CHAPTER 5

HIGHER ORDER INVESTIGATIONS FOR TESTING THEORY IN TIME SERIES ANALYSIS

Let $\{X_t\}$ be a Gaussian ARMA process with spectral density $f_\theta(\lambda)$, where θ is an unknown parameter. The problem considered is that of testing a simple hypothesis $H : \theta = \theta_0$ against the alternative $A : \theta \neq \theta_0$. For this problem we propose a class of tests $\mathbf{S}$, which contains the likelihood ratio (LR), Wald (W), modified Wald (MW) and Rao (R) tests as special cases. Then we derive the χ^2 type asymptotic expansion of the distribution of $T \in \mathbf{S}$ up to order n^{-1}, where n is the sample size. We also derive the χ^2 type asymptotic expansion of the distribution of T under the sequence of alternatives $A_n : \theta = \theta_0 + \epsilon/\sqrt{n}$, $\epsilon > 0$. Then we compare the local powers of the LR, W, MW and R tests on the basis of their asymptotic expansions.

5.1. Asymptotic expansions of the distributions of a class of tests under the null hypothesis

In multivariate analysis, the asymptotic expansions of the distributions of various test statistics have been investigated in detail (e.g., Peers(1971), Hayakawa(1975, 1977), Hayakawa and Puri(1985)). On the other hand, in time series analysis, the first systematic study was tried by Whittle(1951). For an autoregressive process or a moving average process, he gave the limiting distribution of a test statistic of likelihood ratio type, and indicated a method to give its Edgeworth expansion. Recently Phillips(1977) gave the Edgeworth expansion of the t-ratio test statistic in the estimation of the coefficient of a first-order autoregressive process (AR(1)). For an AR(1) process, Tanaka(1982) gave the higher order approximations for the distributions of the likelihood ratio, Wald and Lagrange Multiplier tests under both the null and alternative hypotheses.

In this section we consider a Gaussian ARMA process $\{X_t\}$ with the spectral density $f_\theta(\lambda)$ which depends on an unknown parameter θ. We assume that θ is scalar in order to avoid unnecessarily complex notations and formulas. The problem considered is that of testing a simple hypothesis $H : \theta = \theta_0$ against the alternative $A : \theta \neq \theta_0$. For this problem we propose a class of tests $\mathbf{S}$, which contains the likelihood ratio (LR), Wald (W), modified Wald (MW) and Rao (R) tests as special cases. Then we derive the χ^2 type asymptotic expansion of the distribution of $T \in \mathbf{S}$ up to order $1/n$, where n is the sample size. Also we elucidate a correction factor ρ which makes the term of order $1/n$ in the asymptotic expansion of the distribution of $(1 + \rho/n)T$ vanish (i.e., Bartlett's adjustment) and give the necessary and sufficient condition for $T \in \mathbf{S}$ such that T is adjustable in the sense of Bartlett.

Suppose that the process $\{X_t\}$ satisfies Assumptions 4.1.1 – 4.1.3. Henceforth we use the same notations as those of Chapter 4.

Consider the transformation

$$W_1 = \frac{Z_1}{\sqrt{I}},$$
$$W_2 = Z_2 - J \cdot I^{-1} Z_1,$$
$$W_3 = Z_3 - L \cdot I^{-1} Z_1.$$

For the testing problem $H : \theta = \theta_0$ against $A : \theta \neq \theta_0$, we introduce the following class of tests:

$$\mathbf{S}_H = \left\{ T \middle| T = W_1^2 + \frac{1}{\sqrt{n}}(a_1 W_1^2 W_2 + a_2 W_1^3) + \frac{1}{n}(b_1 W_1^2 + b_2 W_1^2 W_2^2 + b_3 W_1^4 + b_4 W_1^3 W_2 \right.$$
$$\left. + b_5 W_1^3 W_3) + o_p(n^{-1}), \quad \text{under } H, \quad \text{where } a_i \ (i = 1, 2) \text{ and } b_i \ (i = 1, \dots, 5) \right.$$
$$\left. \text{are nonrandom constants} \right\}.$$

This class $\mathbf{S}_H$ is a very natural one.

(i) The likelihood ratio test $LR = 2[L_n(\mathbf{X}_n, \hat{\theta}_{ML}^{(n)}) - L_n(\mathbf{X}_n, \theta_0)]$ belongs to $\mathbf{S}_H$. In fact, expanding LR in a Taylor series at $\theta = \theta_0$, and noting Theorem 2.2.1 and (3.2.4) we obtain

$$LR = 2\left(\hat{\theta}_{ML}^{(n)} - \theta_0\right) \frac{\partial}{\partial\theta} L_n(\mathbf{X}_n, \theta_0) + \left(\hat{\theta}_{ML}^{(n)} - \theta_0\right)^2 \frac{\partial^2}{\partial\theta^2} L_n(\mathbf{X}_n, \theta_0)$$
$$+ \frac{1}{3}\left(\hat{\theta}_{ML}^{(n)} - \theta_0\right)^3 \frac{\partial^3}{\partial\theta^3} L_n(\mathbf{X}_n, \theta_0) + \frac{1}{12}\left(\hat{\theta}_{ML}^{(n)} - \theta_0\right)^4 \frac{\partial^4}{\partial\theta^4} L_n(\mathbf{X}_n, \theta_0) + o_p(n^{-1})$$
$$= 2\sqrt{n}\left(\hat{\theta}_{ML}^{(n)} - \theta_0\right) Z_1(\theta_0) - \left\{\sqrt{n}\left(\hat{\theta}_{ML}^{(n)} - \theta_0\right)\right\}^2 \left\{I(\theta_0) + \frac{\Delta(\theta_0)}{n}\right\}$$
$$+ \frac{1}{\sqrt{n}} Z_2(\theta_0) \left\{\sqrt{n}\left(\hat{\theta}_{ML}^{(n)} - \theta_0\right)\right\}^2 + \frac{1}{3\sqrt{n}} \left\{\sqrt{n}\left(\hat{\theta}_{ML}^{(n)} - \theta_0\right)\right\}^3 \cdot E\frac{1}{n}\frac{\partial^3}{\partial\theta^3} L_n(\mathbf{X}_n, \theta_0)$$
$$+ \frac{1}{3n} \left\{\sqrt{n}\left(\hat{\theta}_{ML}^{(n)} - \theta_0\right)\right\}^3 Z_3(\theta_0)$$
$$+ \frac{1}{12n} \left\{\sqrt{n}\left(\hat{\theta}_{ML}^{(n)} - \theta_0\right)\right\}^4 \cdot E\frac{1}{n}\frac{\partial^4}{\partial\theta^4} L_n(\mathbf{X}_n, \theta_0) + o_p(n^{-1}), \qquad (5.1.1)$$

where $EZ_1(\theta)^2 = I(\theta) + \Delta(\theta)/n + o(n^{-1})$. In view of Theorem 2.3.3, (2.3.22) and (2.3.23) we have

$$LR = W_1^2 + \frac{1}{3\sqrt{n}I^{3/2}}\left(3I^{\frac{1}{2}}W_1^2 W_2 - K W_1^3\right) + \frac{1}{12nI^3}\left[-12I^2 \Delta W_1^2 \right.$$
$$+ 12I W_1^2 W_2^2 + \left\{3(J+K)^2 - I(3M + 6N + H)\right\} W_1^4$$
$$\left. - 12I^{\frac{1}{2}}(J+K)W_1^3 W_2 + 4I^{\frac{3}{2}}W_1^3 W_3\right] + o_p(n^{-1}),$$

which implies that LR belongs to $\mathbf{S}_H$.

Similarly, we can get results (ii) – (iv):

(ii) Wald's test $W = n(\hat{\theta}_{ML}^{(n)} - \theta_0)^2 I(\hat{\theta}_{ML}^{(n)})$ belongs to $\mathbf{S}_H$ with the coefficients $a_1 = 2/I$, $a_2 = J/I^{3/2}$, $b_1 = -2\Delta/I$, $b_2 = 3/I^2$, $b_3 = -(3J^2 + 4JK + K^2)/4I^3 + (4L + 3N + H)/6I^2$, $b_4 = -K/I^{5/2}$ and $b_5 = 1/I^{3/2}$.

(iii)   A modified Wald's test $MW = n(\hat{\theta}_{ML}^{(n)} - \theta_0)^2 I(\theta_0)$ belongs to $\mathbf{S}_H$ with the coefficients $a_1 = 2/I$, $a_2 = -(J+K)/I^{3/2}$, $b_1 = -2\Delta/I$, $b_2 = 3/I^2$, $b_3 = (9J^2 + 14JK + 5K^2)/4I^3 - (L + 3M + 6N + H)/3I^2$, $b_4 = -(6J + 4K)/I^{5/2}$ and $b_5 = 1/I^{3/2}$.

(iv)   Rao's test $R = Z_1(\theta_0)^2 I(\theta_0)^{-1}$ belongs to $\mathbf{S}_H$ with the coefficients $a_1 = a_2 = b_1 = b_2 = b_3 = b_4 = b_5 = 0$.

To derive the asymptotic expansion of the distribution of $T \in \mathbf{S}_H$, we remember the third-order Edgeworth expansion for $\mathbf{W} = (W_1, W_2, W_3)'$ given by (3.2.79):

$$P_{\theta_0}^n[\mathbf{W} \in \mathbf{B}] = \int_{\mathbf{B}} q_n(\mathbf{w}) \, d\mathbf{w} + o(n^{-1}), \tag{5.1.2}$$

where $\mathbf{B}$ is a Borel set of R^3 and $\mathbf{w} = (w_1, w_2, w_3)'$. For $T \in \mathbf{S}_H$, define $c_T(t) = E[e^{itT}]$. It follows from (5.1.2) that

$$
\begin{aligned}
c_T(t) &= \iiint \exp it \left\{ w_1^2 + \frac{1}{\sqrt{n}}(a_1 w_1^2 w_2 + a_2 w_1^3) \right. \\
&\qquad \left. + \frac{1}{n}(b_1 w_1^2 + b_2 w_1^2 w_2^2 + b_3 w_1^4 + b_4 w_1^3 w_2 + b_5 w_1^3 w_3) \right\} q_n(\mathbf{w}) \, d\mathbf{w} + o(n^{-1}) \\
&= \iiint \exp(itw_1^2) \times \left[1 + \frac{it}{\sqrt{n}}(a_1 w_1^2 w_2 + a_2 w_1^3) \right. \\
&\qquad + \frac{it}{n}(b_1 w_1^2 + b_2 w_1^2 w_2^2 + b_3 w_1^4 + b_4 w_1^3 w_2 + b_5 w_1^3 w_3) \\
&\qquad \left. + \frac{(it)^2}{2n}(a_1 w_1^2 w_2 + a_2 w_1^3)^2 \right] q_n(\mathbf{w}) \, d\mathbf{w} + o(n^{-1}).
\end{aligned}
$$

In the first place we calculate the above integral with respect to w_2 and w_3. Second, integrating it with respect to w_1, it is not difficult to show the following lemma.

Lemma 5.1.1. *Under Assumptions 4.1.1 – 4.1.3, the characteristic function $c_T(t)$ has the asymptotic expansion:*

$$c_T(t) = (1 - 2it)^{-\frac{1}{2}} \left[1 + n^{-1} \sum_{j=0}^{3} A_j^{(T)} (1 - 2it)^{-j} \right] + o(n^{-1}),$$

where

$$
\begin{aligned}
A_0^{(T)} &= \left\{ 9I^2(IM - J^2)a_1^2 + 6I(IN - JK)a_1 - 12I^3 b_1 - 12I^2(IM - J^2)b_2 - 12I^2\Delta \right. \\
&\qquad \left. + 3IH - 5K^2 \right\} \big/ 24I^3, \\
A_1^{(T)} &= \left\{ -6I^2(IM - J^2)a_1^2 - 8I(IN - JK)a_1 + 15I^3 a_2^2 + 6I^{\frac{3}{2}}Ka_2 + 4I^3 b_1 \right. \\
&\qquad \left. + 4I^2(IM - J^2)b_2 - 12I^3 b_3 + 4I^2\Delta - 2IH + 5K^2 \right\} \big/ 8I^3, \\
A_2^{(T)} &= \left\{ 3(I^3 M - I^2 J^2)a_1^2 + 6I(IN - JK)a_1 - 30I^3 a_2^2 - 16KI^{\frac{3}{2}}a_2 + 12I^3 b_3 \right. \\
&\qquad \left. + IH - 5K^2 \right\} \big/ 8I^3, \\
A_3^{(T)} &= 5\left(3I^{\frac{3}{2}}a_2 + K\right)^2 \big/ 24I^3.
\end{aligned}
$$

From the above lemma we have

Theorem 5.1.1. *Under Assumptions 4.1.1 – 4.1.3, the asymptotic expansion of the distribution of $T \in \mathbf{S}_H$ is given by*

$$P^n_{\theta_0}[T \leq x] = P[\chi_1^2 \leq x] + n^{-1} \sum_{j=0}^{3} A_j^{(T)} P[\chi_{1+2j}^2 \leq x] + o(n^{-1}). \tag{5.1.3}$$

For concrete spectral models we can give the coefficients $A_j^{(T)}$ in (5.1.3) for the four tests T = LR, W, MW and R in simple forms (cf. Proposition 2.3.1).

Example 5.1.1. For the autoregressive spectral density

$$f_{\theta_0}(\lambda) = \frac{\sigma^2}{2\pi}|1 - \alpha e^{i\lambda}|^{-2} \quad (\theta_0 = \alpha),$$

we can show that
(i) for $T = LR$ (likelihood ratio test),

$$A_0^{(LR)} = 1, \ A_1^{(LR)} = -1, \ A_2^{(LR)} = A_3^{(LR)} = 0;$$

(ii) for $T = W$ (Wald's test),

$$A_0^{(W)} = \frac{5\alpha^2 - 1}{4(1 - \alpha^2)}, \ A_1^{(W)} = -\frac{\alpha^2 + 1}{2(1 - \alpha^2)}, \ A_2^{(W)} = \frac{3}{4}, \ A_3^{(W)} = 0;$$

(iii) for $T = MW$ (modified Wald's test),

$$A_0^{(MW)} = \frac{5\alpha^2 - 1}{4(1 - \alpha^2)}, \ A_1^{(MW)} = \frac{2 - \alpha^2}{2(1 - \alpha^2)}, \ A_2^{(MW)} = \frac{-33\alpha^2 - 3}{4(1 - \alpha^2)}, \ A_3^{(MW)} = \frac{15\alpha^2}{2(1 - \alpha^2)};$$

(iv) for $T = R$ (Rao's test),

$$A_0^{(R)} = \frac{11 - 15\alpha^2}{4(1 - \alpha^2)}, \ A_1^{(R)} = \frac{27\alpha^2 - 10}{2(1 - \alpha^2)}, \ A_2^{(R)} = \frac{9 - 69\alpha^2}{4(1 - \alpha^2)}, \ A_3^{(R)} = \frac{15\alpha^2}{2(1 - \alpha^2)}.$$

Example 5.1.2. For the moving average spectral density

$$f_{\theta_0}(\lambda) = \frac{\sigma^2}{2\pi}|1 - \beta e^{i\lambda}|^2 \quad (\theta_0 = \beta),$$

we can show that
(i) for $T = LR$,

$$A_0^{(LR)} = -\frac{1 + 2\beta^2}{2(1 - \beta^2)}, \ A_1^{(LR)} = \frac{1 + 2\beta^2}{2(1 - \beta^2)}, \ A_2^{(LR)} = A_3^{(LR)} = 0;$$

(ii) for $T = W$,

$$A_0^{(W)} = \frac{-9 - 7\beta^2}{4(1 - \beta^2)}, \ A_1^{(W)} = \frac{5\beta^2 - 3}{2(1 - \beta^2)}, \ A_2^{(W)} = \frac{15 - 33\beta^2}{4(1 - \beta^2)}, \ A_3^{(W)} = \frac{15\beta^2}{2(1 - \beta^2)};$$

(iii) for $T = MW$,

$$A_0^{(MW)} = \frac{-9 - 7\beta^2}{4(1 - \beta^2)}, \quad A_1^{(MW)} = \frac{5\beta^2}{2(1 - \beta^2)}, \quad A_2^{(MW)} = \frac{-3\beta^2 + 9}{4(1 - \beta^2)}, \quad A_3^{(MW)} = 0;$$

(iv) for $T = R$,

$$A_0^{(R)} = \frac{11 - 3\beta^2}{4(1 - \beta^2)}, \quad A_1^{(R)} = \frac{21\beta^2 - 10}{2(1 - \beta^2)}, \quad A_2^{(R)} = \frac{3(3 - 23\beta^2)}{2(1 - \beta^2)}, \quad A_3^{(R)} = \frac{15\beta^2}{2(1 - \beta^2)}.$$

We next elucidate Bartlett's adjustment for $T \in \mathbf{S}_H$. Since $T \in \mathbf{S}_H$, it is easy to show that

$$E(T) = 1 - \frac{\rho}{n} + o(n^{-1}),$$

where

$$\rho = -\frac{I^2\Delta + I^3 b_1 + I^2(IM - J^2)b_2 + 3I^3 b_3 + Ia_1(IN - JK) + I^{3/2}Ka_2}{I^3}.$$

Thus we have

$$\frac{T}{E(T)} = \left(1 + \frac{\rho}{n}\right)T + o_p(n^{-1}).$$

The above ρ is called Bartlett's adjustment factor. If the terms of order n^{-1} in the asymptotic expansion of the distribution of $T^* = (1 + \frac{\rho}{n})T$ vanish (i.e., $P_{\theta_0}^n[T^* \leq x] = P[\chi_1^2 \leq x] + o(n^{-1})$), we say that T is adjustable in the sense of Bartlett.

Denoting $c_{T^*}(t) = Ee^{itT^*}$, we have

$$
\begin{aligned}
c_{T^*}(t) &= c_T(t) + E\left\{e^{itW_1^2} \cdot \frac{it\rho W_1^2}{n}\right\} + o(n^{-1}) \\
&= c_T(t) + (1 - 2it)^{-\frac{1}{2}}\left\{\frac{\rho}{2n}\left(\frac{1}{1 - 2it} - 1\right)\right\} + o(n^{-1}) \\
&= (1 - 2it)^{-\frac{1}{2}}\left[1 + n^{-1}\left\{A_0^{(T)} - \frac{\rho}{2} + \left(A_1^{(T)} + \frac{\rho}{2}\right)(1 - 2it)^{-1}\right.\right. \\
&\quad \left.\left. + A_2^{(T)}(1 - 2it)^{-2} + A_3^{(T)}(1 - 2it)^{-3}\right\}\right] + o(n^{-1}).
\end{aligned}
$$

(5.1.4)

In (5.1.4), putting $A_0^{(T)} - \rho/2 = 0$, $A_1^{(T)} + \rho/2 = 0$, $A_2^{(T)} = 0$ and $A_3^{(T)} = 0$, we have the following theorem.

Theorem 5.1.2. *The test statistic $T \in \mathbf{S}_H$ is adjustable in the sense of Bartlett if and only if the coefficients $\{a_j\}$ and $\{b_j\}$ satisfy the relations (i) and (ii):*

(i) $a_2 = -\dfrac{K}{3I^{3/2}}$,

(ii) $3I^2(IM - J^2)a_1^2 + 6I(IN - JK)a_1 + 12I^3 b_3 + IH - 3K^2 = 0.$

Among the four tests LR, W, MW and R, the LR test is the only one which is adjustable in the sense of Bartlett.

For the LR test, Bartlett's adjustment factor $\rho = \rho_{LR}(\theta_0)$ is given by

$$\rho_{LR}(\theta_0) = \frac{-M + 2N + H}{4I^2} + \frac{3J^2 - 6JK - 5K^2}{12I^3}.$$

In particular, for the ARMA spectral density

$$f_{\theta_0}(\lambda) = \frac{\sigma^2}{2\pi} \frac{|1 - \beta e^{i\lambda}|^2}{|1 - \alpha e^{i\lambda}|^2}$$

Bartlett's adjustment factors are given by

$$\rho_{LR}(\sigma^2) = -\frac{1}{3} \quad \text{for} \quad \theta_0 = \sigma^2,$$

$$\rho_{LR}(\alpha) = 2 \quad \text{for} \quad \theta_0 = \alpha,$$

$$\rho_{LR}(\beta) = \frac{-1 - 2\beta^2}{1 - \beta^2} \quad \text{for} \quad \theta_0 = \beta.$$

5.2. Comparisons of powers of a class of tests under a local alternative

In this section we introduce a class $\mathbf{S}_A$ of tests and derive the χ^2 type asymptotic expansion of the distribution of $S \in \mathbf{S}_A$ under the sequence of alternatives $A_n : \theta = \theta_0 + \epsilon/\sqrt{n}$, $\epsilon > 0$. Using the asymptotic expansion for S, we compare the local powers of the LR, W, MW and R tests on the basis of their asymptotic expansions. Then it is shown that none of the above tests is uniformly superior. However, if we modify them to be asymptotically unbiased we can show that their local powers are identical.

Consider the transformation

$$U_1(\theta) = \frac{Z_1(\theta)}{\sqrt{I(\theta)}},$$

$$U_2(\theta) = \frac{Z_2(\theta) - J(\theta)I(\theta)^{-1}Z_1(\theta)}{\gamma_\theta I(\theta)},$$

where $\gamma_\theta = \{M(\theta)I(\theta) - J(\theta)^2\}^{1/2}/I(\theta)^{3/2}$. In this section, for simplicity, we use U_1, U_2, Z_1, Z_2, I, J, K, γ, instead of $U_1(\theta)$, $U_2(\theta)$, $Z_1(\theta)$, $Z_2(\theta)$, $I(\theta)$, $J(\theta)$, $K(\theta)$, γ_θ, respectively, if they are evaluated at $\theta = \theta_0 + \epsilon/\sqrt{n}$. Define the following class of tests:

$$\begin{aligned}
\mathbf{S}_A = \Bigg\{ S \Big| S &= \left\{U_1 + I(\theta_0)^{\frac{1}{2}}\epsilon\right\}^2 + \frac{1}{\sqrt{n}} \Big[c_1 U_1^3 + c_2 U_1^2 U_2 \\
&\quad + \{c_3 U_1^2 + c_4 U_1 U_2\}\epsilon + \{c_5 U_1 + c_6 U_2\}\epsilon^2 + c_7\epsilon^3 \Big] + o_p(n^{-\frac{1}{2}}), \\
&\quad \text{under } A_n, \quad \text{where } c_7 = I^{\frac{3}{2}}c_1 - Ic_3 + I^{\frac{1}{2}}c_5 \Bigg\},
\end{aligned}$$

This class $\mathbf{S}_A$ is also very natural.

(i) The likelihood ratio test $LR = 2[L_n(\mathbf{X}_n, \hat{\theta}_{ML}^{(n)}) - L_n(\mathbf{X}_n, \theta_0)]$ belongs to $\mathbf{S}_A$. In fact, expanding LR in a Taylor series at $\theta = \hat{\theta}_{ML}^{(n)}$, we obtain

$$LR = -\left(\theta_0 - \hat{\theta}_{ML}^{(n)}\right)^2 \frac{\partial^2}{\partial\theta^2} L_n\left(\mathbf{X}_n, \hat{\theta}_{ML}^{(n)}\right) + \frac{1}{3}\left(\hat{\theta}_{ML}^{(n)} - \theta_0\right)^3 \frac{\partial^3}{\partial\theta^3} L_n\left(\mathbf{X}_n, \hat{\theta}_{ML}^{(n)}\right) + o_p(n^{-\frac{1}{2}})$$

$$
\begin{aligned}
&= -\left(\hat{\theta}_{ML}^{(n)} - \theta + \theta - \theta_0\right)^2 \left\{\frac{\partial^2}{\partial\theta^2} L_n(\mathbf{X}_n, \theta) + \left(\hat{\theta}_{ML}^{(n)} - \theta\right) \frac{\partial^3}{\partial\theta^3} L_n(\mathbf{X}_n, \theta)\right\} \\
&\quad + \frac{1}{3}\left(\hat{\theta}_{ML}^{(n)} - \theta + \theta - \theta_0\right)^3 \frac{\partial^3}{\partial\theta^3} L_n(\mathbf{X}_n, \theta) + o_p(n^{-\frac{1}{2}}) \\
&= -\frac{1}{n}\frac{\partial^2}{\partial\theta^2} L_n(\mathbf{X}_n, \theta)(v^2 + 2v\epsilon + \epsilon^2) \\
&\quad - \frac{1}{3\sqrt{n}}\left\{\frac{1}{n}\frac{\partial^3}{\partial\theta^3} L_n(\mathbf{X}_n, \theta)\right\}(2v^3 + 3v^2\epsilon - \epsilon^3) + o_p(n^{-\frac{1}{2}}),
\end{aligned} \tag{5.2.1}
$$

where $v = \sqrt{n}(\hat{\theta}_{ML}^{(n)} - \theta)$. In view of Theorem 2.3.3 and (2.3.22) we can see that

$$
\begin{aligned}
LR &= \left\{U_1 + I(\theta_0)^{\frac{1}{2}}\epsilon\right\}^2 + \frac{1}{\sqrt{n}}\left[-\frac{K}{3I^{3/2}}U_1^3 + \gamma U_1^2 U_2 \right. \\
&\quad \left. + \left\{\frac{J+K}{I^{1/2}}U_1 - \gamma I U_2\right\}\epsilon^2 + \frac{3J+2K}{3}\epsilon^3\right] + o_p(n^{-\frac{1}{2}}).
\end{aligned}
$$

Similarly, we can get results (ii) – (iv):

(ii) Wald's test $W = n(\hat{\theta}_{ML}^{(n)} - \theta_0)^2 I(\hat{\theta}_{ML}^{(n)})$ belongs to $\mathbf{S}_A$ with the coefficients $c_1 = J/I^{3/2}$, $c_2 = 2\gamma$, $c_3 = (3J+K)/I$, $c_4 = 2\gamma I^{1/2}$, $c_5 = 2(2J+K)/I^{1/2}$, $c_6 = 0$ and $c_7 = 2J+K$.

(iii) The modified Wald's test $MW = n(\hat{\theta}_{ML}^{(n)} - \theta_0)^2 I(\theta_0)$ belongs to $\mathbf{S}_A$ with the coefficients $c_1 = -(J+K)/I^{3/2}$, $c_2 = 2\gamma$, $c_3 = -(3J+2K)/I$, $c_4 = 2\gamma I^{1/2}$, $c_5 = -(2J+K)/I^{1/2}$ and $c_6 = c_7 = 0$.

(iv) Rao's test $R = Z_1(\theta_0)^2 I(\theta_0)^{-1}$ belongs to $\mathbf{S}_A$ with the coefficients $c_1 = c_2 = 0$, $c_3 = K/I$, $c_4 = -2\gamma I^{1/2}$, $c_5 = (J+2K)/I^{1/2}$, $c_6 = -2\gamma I$ and $c_7 = J+K$.

The following lemma is shown by a slight modification for Lemma 3.2.9.

Lemma 5.2.1. *Under Assumptions 4.1.1 – 4.1.3,*

$$
\begin{aligned}
&P_{\theta_0 + \epsilon/\sqrt{n}}^n[U_1 < y_1, \ U_2 < y_2] \\
&= \int_{-\infty}^{y_1}\int_{-\infty}^{y_2} \phi(u_1)\phi(u_2) \times \left[1 + \frac{1}{6\sqrt{n}}\left\{\frac{K(\theta_0)}{I(\theta)_0^{3/2}}(u_1^3 - 3u_1) + 3c_{112}(u_1^2 u_2 - u_2)\right.\right. \\
&\quad \left.\left. + 3c_{122}(u_1 u_2^2 - u_1) + c_{222}(u_2^3 - 3u_2)\right\}\right] du_1 du_2 + o(n^{-\frac{1}{2}}) \\
&= \int_{-\infty}^{y_1}\int_{-\infty}^{y_2} f(u_1, u_2) du_1 du_2 + o(n^{-\frac{1}{2}}) \quad \text{(say)},
\end{aligned}
$$

where $\phi(u) = \{2\pi\}^{-1/2}\exp(-u^2/2)$, and the coefficients c_{112}, c_{122} and c_{222} are expressed by the spectral density.

Using Lemma 5.2.1 we can evaluate the characteristic function $c_S(t)$ of $S \in \mathbf{S}_A$, under A_n. In fact,

$$
c_S(t) = E_{\theta_0 + \epsilon/\sqrt{n}}\{e^{itS}\}
$$

$$
= \iint f(u_1, u_2) \exp \left[it \left\{ u_1 + I(\theta_0)^{1/2} \epsilon \right\}^2 \right]
$$

$$
\times \left[1 + \frac{it}{\sqrt{n}} \left\{ c_1 u_1^3 + c_2 u_1^2 u_2 + (c_3 u_1^2 + c_4 u_1 u_2) \epsilon \right. \right.
$$

$$
\left. \left. + (c_5 u_1 + c_6 u_2) \epsilon^2 + c_7 \epsilon^3 \right\} \right] du_1 du_2 + o(n^{-\frac{1}{2}})
$$

$$
= \iint \phi(u_1) \phi(u_2) \exp \left[it \left\{ u_1 + I(\theta_0)^{1/2} \epsilon \right\}^2 \right]
$$

$$
\times \left[1 + \frac{it}{\sqrt{n}} \left\{ c_1 u_1^3 + c_2 u_1^2 u_2 + (c_3 u_1^2 + c_4 u_1 u_2) \epsilon + (c_5 u_1 + c_6 u_2) \epsilon^2 + c_7 \epsilon^3 \right\} \right.
$$

$$
+ \frac{1}{6\sqrt{n}} \left\{ \frac{K(\theta_0)}{I(\theta_0)^{3/2}} (u_1^3 - 3u_1) + 3c_{112}(u_1^2 u_2 - u_2) \right.
$$

$$
\left. \left. + 3c_{122}(u_1 u_2^2 - u_1) + c_{222}(u_2^3 - 3u_2) \right\} \right] du_1 du_2 + o(n^{-\frac{1}{2}}). \tag{5.2.2}
$$

Integration of (5.2.2) with respect to u_2 yields

$$
c_S(t) = \exp \left\{ \frac{it I(\theta_0) \epsilon^2}{1 - 2it} \right\} \cdot (1 - 2it)^{-\frac{1}{2}} \int (2\pi)^{-\frac{1}{2}} (1 - 2it)^{\frac{1}{2}}
$$

$$
\times \exp \left[-\frac{1 - 2it}{2} \left\{ u_1 - \frac{2\epsilon it I(\theta_0)^{1/2}}{1 - 2it} \right\}^2 \right]
$$

$$
\times \left[1 + \frac{it}{\sqrt{n}} \left\{ c_1 u_1^3 + c_3 u_1^2 \epsilon + c_5 u_1 \epsilon^2 + c_7 \epsilon^3 \right\} \right.
$$

$$
\left. + \frac{K(\theta_0)}{6\sqrt{n} I(\theta_0)^{3/2}} (u_1^3 - 3u_1) \right] du_1 + o(n^{-\frac{1}{2}}).
$$

Calculation of the above integral leads to

Lemma 5.2.2. *Under Assumptions 4.1.1 – 4.1.3, the characteristic function $c_S(t)$ of $S \in \mathbf{S}_A$ under $\theta = \theta_0 + \epsilon/\sqrt{n}$ has the asymptotic expansion*

$$
c_S(t) = \exp \left\{ \frac{it I(\theta_0) \epsilon^2}{1 - 2it} \right\} \times (1 - 2it)^{-\frac{1}{2}} \left[1 + n^{-\frac{1}{2}} \sum_{j=0}^{3} B_j^{(S)} (1 - 2it)^{-j} \right] + o(n^{-\frac{1}{2}}),
$$

where

$$
B_0^{(S)} = \frac{1}{6} \left[\left\{ -9I(\theta_0)^{\frac{3}{2}} c_1 + 6I(\theta_0) c_3 - 3I(\theta_0)^{\frac{1}{2}} c_5 - K(\theta_0) \right\} \epsilon^3 \right.
$$

$$
\left. + \left\{ 9I(\theta_0)^{\frac{1}{2}} c_1 - 3c_3 + \frac{3K(\theta_0)}{I(\theta_0)} \right\} \epsilon \right],
$$

$$
B_1^{(S)} = \frac{1}{2} \left[\left\{ 6I(\theta_0)^{\frac{3}{2}} c_1 - 3I(\theta_0) c_3 + I(\theta_0)^{\frac{1}{2}} c_5 + K(\theta_0) \right\} \epsilon^3 \right.
$$

$$
\left. + \left\{ c_3 - 6I(\theta_0)^{\frac{1}{2}} c_1 - \frac{2K(\theta_0)}{I(\theta_0)} \right\} \epsilon \right],
$$

$$
B_2^{(S)} = \frac{1}{2} \left[\left\{ I(\theta_0) c_3 - 4I(\theta_0)^{\frac{3}{2}} c_1 - K(\theta_0) \right\} \epsilon^3 + \left\{ 3I(\theta_0)^{\frac{1}{2}} c_1 + \frac{K(\theta_0)}{I(\theta_0)} \right\} \epsilon \right],
$$

$$B_3^{(S)} = \frac{1}{6}\left\{3I(\theta_0)^{\frac{3}{2}}c_1 + K(\theta_0)\right\}\epsilon^3.$$

Lemma 5.2.2 implies

Theorem 5.2.1. *Under Assumptions 4.1.1 – 4.1.3, the distribution function of $S \in \mathbf{S}_A$ for $\theta = \theta_0 + \epsilon/\sqrt{n}$ has the asymptotic expansion*

$$P_{\theta_0+\epsilon/\sqrt{n}}^n[S \le x] = P[\chi_1^2(\delta) \le x] + n^{-\frac{1}{2}}\sum_{j=0}^{3}B_j^{(S)}P[\chi_{1+2j}^2(\delta) \le x] + o(n^{-\frac{1}{2}}),$$

where $\delta^2 = I(\theta_0)\epsilon^2/2$, and $\chi_j^2(\delta)$ is a noncentral χ^2 random variable with j degrees of freedom and noncentrality parameter δ^2.

For the four tests $S = LR$, W, MW and R, we can give more explicit expressions for the coefficients $B_j^{(S)}$ in Theorem 5.2.1.

Example 5.2.1.
(i) $S = LR$ (likelihood ratio test)

$$B_0^{(LR)} = -\frac{(3J(\theta_0) + K(\theta_0))\epsilon^3}{6}, \quad B_1^{(LR)} = \frac{J(\theta_0)\epsilon^3}{2},$$

$$B_2^{(LR)} = \frac{K(\theta_0)\epsilon^3}{6}, \quad B_3^{(LR)} = 0,$$

(ii) $S = W$ (Wald's test)

$$B_0^{(W)} = -\frac{(K(\theta_0) + 3J(\theta_0))\epsilon^3}{6},$$

$$B_1^{(W)} = \frac{J(\theta_0)\epsilon^3 - (3J(\theta_0) + K(\theta_0))\epsilon/I(\theta_0)}{2},$$

$$B_2^{(W)} = \frac{-J(\theta_0)\epsilon^3 + (3J(\theta_0) + K(\theta_0))\epsilon/I(\theta_0)}{2},$$

$$B_3^{(W)} = \frac{(K(\theta_0) + 3J(\theta_0))\epsilon^3}{6},$$

(iii) $S = MW$ (modified Wald's test)

$$B_0^{(MW)} = -\frac{(K(\theta_0) + 3J(\theta_0))\epsilon^3}{6},$$

$$B_1^{(MW)} = \frac{J(\theta_0)\epsilon^3 + (3J(\theta_0) + 2K(\theta_0))\epsilon/I(\theta_0)}{2},$$

$$B_2^{(MW)} = \frac{(K(\theta_0) + J(\theta_0))\epsilon^3 - (3J(\theta_0) + 2K(\theta_0))\epsilon/I(\theta_0)}{2},$$

$$B_3^{(MW)} = -\frac{(2K(\theta_0) + 3J(\theta_0))\epsilon^3}{6},$$

(iv) $S = R$ (Rao's test)

$$B_0^{(R)} = -\frac{(K(\theta_0) + 3J(\theta_0))\epsilon^3}{6}, \quad B_1^{(R)} = \frac{J(\theta_0)\epsilon^3 - K(\theta_0)\epsilon/I(\theta_0)}{2},$$
$$B_2^{(R)} = \frac{K(\theta_0)\epsilon}{2I(\theta_0)}, \quad B_3^{(R)} = \frac{K(\theta_0)\epsilon^3}{6}.$$

In view of Theorem 5.2.1 we can investigate the local power properties in the class $\mathbf{S}_A$. By Theorem 5.2.1 and Example 5.2.1, it is not difficult to show that for $S \in \mathbf{S}_A$,

$$
\begin{aligned}
&P_{\theta_0+\epsilon/\sqrt{n}}^n[S > x] - P_{\theta_0+\epsilon/\sqrt{n}}^n[LR > x] \\
&= \frac{1}{\sqrt{n}} \left[\frac{1}{2} \left\{ P\left(\chi_7^2(\delta) > x\right) - P\left(\chi_5^2(\delta) > x\right) \right\} Q_3^{(S)}(\theta_0) \right. \\
&\quad + \frac{1}{2} \left\{ P\left(\chi_5^2(\delta) > x\right) - P\left(\chi_3^2(\delta) > x\right) \right\} Q_2^{(S)}(\theta_0) \\
&\quad \left. + \frac{1}{2} \left\{ P\left(\chi_3^2(\delta) > x\right) - P\left(\chi_1^2(\delta) > x\right) \right\} Q_1^{(S)}(\theta_0) \right] + o(n^{-\frac{1}{2}}),
\end{aligned} \tag{5.2.3}
$$

where

$$
\begin{aligned}
Q_1^{(S)}(\theta_0) &= \left\{ 3I(\theta_0)^{\frac{3}{2}}c_1 - 2I(\theta_0)c_3 + I(\theta_0)^{\frac{1}{2}}c_5 - J(\theta_0) \right\} \epsilon^3 \\
&\quad + \left\{ c_3 - 3I(\theta_0)^{\frac{1}{2}}c_1 - \frac{K(\theta_0)}{I(\theta_0)} \right\} \epsilon, \\
Q_2^{(S)}(\theta_0) &= \left\{ I(\theta_0)c_3 - 3I(\theta_0)^{\frac{3}{2}}c_1 - K(\theta_0) \right\} \epsilon^3 + \left\{ 3I(\theta_0)^{\frac{1}{2}}c_1 + \frac{K(\theta_0)}{I(\theta_0)} \right\} \epsilon, \\
Q_3^{(S)}(\theta_0) &= \frac{\left\{ 3I(\theta_0)^{\frac{3}{2}}c_1 + K(\theta_0) \right\} \epsilon^3}{3}.
\end{aligned}
$$

The following relation is well known,

$$P\left[\chi_{j+2}^2(\delta) > x\right] - P\left[\chi_j^2(\delta) > x\right] = 2p_{j+2}(x; \delta), \tag{5.2.4}$$

where $p_j(x; \delta)$ is the probability density function of $\chi_j^2(\delta)$. (5.2.3) and (5.2.4) above imply

Theorem 5.2.2. *Under Assumptions 4.1.1 – 4.1.3,*

$$
\begin{aligned}
&P_{\theta_0+\epsilon/\sqrt{n}}^n[S > x] - P_{\theta_0+\epsilon/\sqrt{n}}^n[LR > x] \\
&= \frac{1}{\sqrt{n}} \left[Q_3^{(S)}(\theta_0)p_7(x; \delta) + Q_2^{(S)}(\theta_0)p_5(x; \delta) + Q_1^{(S)}(\theta_0)p_3(x; \delta) \right] + o(n^{-\frac{1}{2}}), \quad \text{for } S \in \mathbf{S}_A.
\end{aligned}
$$

By Theorem 5.2.2, for an ARMA process, we can compare the local power properties among the four tests LR, W, MW and R.

Consider the ARMA(p, q) spectral density

$$f_{\theta_0}(\lambda) = \frac{\sigma^2}{2\pi} \frac{\prod_{k=1}^q (1 - \psi_k e^{i\lambda})(1 - \psi_k e^{-i\lambda})}{\prod_{k=1}^p (1 - \rho_k e^{i\lambda})(1 - \rho_k e^{-i\lambda})}, \tag{5.2.5}$$

where $\psi_1, \ldots, \psi_q$, $\rho_1, \ldots, \rho_p$ are real numbers such that $|\psi_j| < 1$, $j = 1, \ldots, q$, $|\rho_j| < 1$, $j = 1, \ldots, p$. For the spectral density (5.2.5) we can get the following local power comparisons.

Example 5.2.2. W versus LR under A_n,

$$P_{\theta_0+\epsilon/\sqrt{n}}^n[W > x] - P_{\theta_0+\epsilon/\sqrt{n}}^n[LR > x]$$
$$= \frac{1}{\sqrt{n}}\{3J(\theta_0) + K(\theta_0)\}\left\{\frac{\epsilon^3}{3}p_7(x;\delta) + \frac{\epsilon}{I(\theta_0)}p_5(x;\delta)\right\} + o(n^{-\frac{1}{2}}).$$

(i) If $\theta_0 = \sigma^2$, then $3J(\theta_0) + K(\theta_0) = -2/\sigma^6 < 0$, which implies that LR is more powerful than W.

(ii) If $\theta_0 = \psi_k$, then $3J(\theta_0) + K(\theta_0) = 6\psi_k/(1 - \psi_k^2)^2$, which implies that W is more powerful than LR if $\psi_k > 0$ and vice versa.

(iii) If $\theta_0 = \rho_k$, then $3J(\theta_0) + K(\theta_0) = 0$, which implies that LR and W have identical local powers.

Example 5.2.3. MW versus LR under A_n,

$$P_{\theta_0+\epsilon/\sqrt{n}}^n[MW > x] - P_{\theta_0+\epsilon/\sqrt{n}}^n[LR > x]$$
$$= \frac{1}{\sqrt{n}}\{-3J(\theta_0) - 2K(\theta_0)\}\left\{\frac{\epsilon^3}{3}p_7(x;\delta) + \frac{\epsilon}{I(\theta_0)}p_5(x;\delta)\right\} + o(n^{-\frac{1}{2}}).$$

(i) If $\theta_0 = \sigma^2$, then $-3J(\theta_0) - 2K(\theta_0) = 1/\sigma^6 > 0$, which implies that MW is more powerful than LR.

(ii) If $\theta_0 = \psi_k$, then $-3J(\theta_0) - 2K(\theta_0) = 0$, which implies that MW and LR have identical local powers.

(iii) If $\theta_0 = \rho_k$, then $-3J(\theta_0) - 2K(\theta_0) = -6\rho_k/(1 - \rho_k^2)^2$, which implies that LR is more powerful than MW if $\rho_k > 0$ and vice versa.

Example 5.2.4. R versus LR under A_n,

$$P_{\theta_0+\epsilon/\sqrt{n}}^n[R > x] - P_{\theta_0+\epsilon/\sqrt{n}}^n[LR > x] = \frac{K(\theta_0)}{\sqrt{n}}\left\{\frac{\epsilon^3}{3}p_7(x;\delta) + \frac{\epsilon}{I(\theta_0)}p_5(x;\delta)\right\} + o(n^{-\frac{1}{2}}).$$

(i) If $\theta_0 = \sigma^2$, then $K(\theta_0) = 1/\sigma^6 > 0$, which implies that R is more powerful than LR.

(ii) If $\theta_0 = \psi_k$, then $K(\theta_0) = -6\psi_k/(1 - \psi_k^2)^2$, which implies that R is more powerful than LR if $\psi_k < 0$ and vice versa.

(iii) If $\theta_0 = \rho_k$, then $K(\theta_0) = 6\rho_k/(1 - \rho_k^2)^2$, which implies that LR is more powerful than R if $\psi_k < 0$ and vice versa.

These examples show that none of the LR, W, MW and R tests is uniformly superior.

Finally we show that an appropriate modification of $S \in \mathbf{S}_A$ leads to a unified result. First, we note that the coefficients c_1, c_3 and c_5 in the stochastic expansions of the four tests automatically satisfy

$$\begin{cases} Ic_3 - 3I^{\frac{1}{2}}c_1 = K \\ I^{\frac{1}{2}}c_5 - Ic_3 = J + K \end{cases} \tag{5.2.6}$$

Henceforth we confine ourselves to a class of tests $\mathbf{S}'_A = \{S|S \in \mathbf{S}_A \text{ and } c_1, c_3 \text{ and } c_5 \text{ satisfy} (5.2.6) \}$.

Furthermore we impose the second-order asymptotic unbiasedness;

$$\frac{\partial}{\partial \epsilon} P^n_{\theta_0 + \epsilon/\sqrt{n}}[S > x]\Big|_{\epsilon=0} = o(n^{-\frac{1}{2}}), \tag{5.2.7}$$

for $S \in \mathbf{S}'_A$. By Lemma 5.2.2 and Theorem 5.2.1, we can see that (5.2.7) is equivalent to

$$\begin{cases} 9I^{\frac{1}{2}}c_1 - 3c_3 + 3K \cdot I^{-1} = 0 \\ c_3 - 6I^{\frac{1}{2}}c_1 - 2K \cdot I^{-1} = 0 \\ 3I^{\frac{1}{2}}c_1 + K \cdot I^{-1} = 0. \end{cases} \tag{5.2.8}$$

Consider a class of tests

$$\mathbf{US}'_A = \{S|S \in \mathbf{S}'_A \text{ and } S \text{ satisfies } (5.2.7) \text{ or } (5.2.8) \}.$$

From Example 5.2.1, it is easy to see $LR \in \mathbf{US}'_A$.
By Theorem 5.2.2 we have

$$P^n_{\theta_0 + \epsilon/\sqrt{n}}[S > x] - P^n_{\theta_0 + \epsilon/\sqrt{n}}[LR > x] = o(n^{-\frac{1}{2}}), \tag{5.2.9}$$

for all $S \in \mathbf{US}'_A$.

Now we modify $S \in \mathbf{S}'_A$ to be second-order asymptotically unbiased. Put $S^\star = m(\hat{\theta}^{(n)}_{ML})S$, where $S \in \mathbf{S}'_A$ and $m(\theta)$ is a continuously twice differentiable function with $m(\theta_0) = 1$. Then

$$\begin{aligned} S^\star &= \left\{ m(\theta) + \left(\hat{\theta}^{(n)}_{ML} - \theta \right) m'(\theta) \right\} S + o_p(n^{-\frac{1}{2}}) \\ &= \left\{ 1 + \frac{\epsilon}{\sqrt{n}} m'(\theta_0) + \frac{U_1}{\sqrt{nI}} m'(\theta_0) \right\} S + o_p(n^{-\frac{1}{2}}) \\ &= \left\{ U_1 + \sqrt{I}\epsilon \right\}^2 + \frac{1}{\sqrt{n}} \left[\left(c_1 + \frac{h}{\sqrt{I}} \right) U_1^3 + c_2 U_1^2 U_2 \right. \\ &\quad + \left\{ (c_3 + 3h)U_1^2 + c_4 U_1 U_2 \right\} \epsilon + \left\{ \left(c_5 + 3\sqrt{I}h \right) U_1 + c_6 U_2 \right\} \epsilon^2 \\ &\quad \left. + (c_7 + Ih)\,\epsilon^3 \right] + o_p(n^{-\frac{1}{2}}), \end{aligned} \tag{5.2.10}$$

where $h = m'(\theta_0)$. Thus we can see that $S^\star \in \mathbf{US}'_A$ if $c_1 + h/\sqrt{I} = -K/(3I^{3/2})$ and $c_3 + 3h = 0$. Summarizing the above we have the following unified result.

Theorem 5.2.3. *Under Assumptions 4.1.1 – 4.1.3, the local powers of all the modified tests $S^\star = m(\hat{\theta}^{(n)}_{ML})S$, $S \in \mathbf{S}'_A$, with $m(\theta_0) = 1$ and $m'(\theta_0) = -\sqrt{I}c_1 - K/(3I)$ (or $= -c_3/3$ equivalently) are identical up to order $n^{-1/2}$.*

CHAPTER 6

HIGHER ORDER ASYMPTOTIC THEORY FOR MULTIVARIATE TIME SERIES

The present chapter studies the higher order asymptotic theory for multivariate time series. Suppose that $\{\mathbf{X}(t)\}$ is a vector-valued Gaussian stationary process with zero mean. For a partial realization $\{\mathbf{X}(1), \ldots, \mathbf{X}(n)\}$, define the sample covariance matrix by $S = \frac{1}{n} \sum_{t=1}^{n} \mathbf{X}(t)\mathbf{X}(t)'$. We are interested in the eigenvalues of S because they play a fundamental role in multivariate analysis. Then the asymptotic expansion for the distribution of functions of the eigenvalues of S will be given. In Section 6.2 we shall derive the asymptotic expansions for certain functions of the sample canonical correlations in multivariate time series. The contents of this chapter are mainly based on Taniguchi and Krishnaiah(1987).

6.1. Asymptotic expansions of the distributions of functions of the eigenvalues of sample covariance matrix in multivariate time series

Recently several authors have introduced some multivariate methods to multivariate time series analysis. Using the orthogonality and asymptotic normality of the "Fourier components" of time series, Priestley *et al.*(1973) and Brillinger(1975) discussed the principal component analysis in the frequency domain, and investigated various limiting properties of certain statistics.

Here we develop a higher order investigation for statistics related to the sample covariance matrix in multivariate time series. Assume that $\{\mathbf{X}(t) = (X_1(t), \ldots, X_p(t))' : t = 0, \pm 1, \ldots\}$ is a $p \times 1$-vector-valued Gaussian stationary process with zero mean and covariance matrix $\Gamma(j) = E\{\mathbf{X}(t)\mathbf{X}(t+j)'\}$.

Assumption 6.1.1. The covariance matrix satisfies

$$\sum_{u=-\infty}^{\infty} |u| \, \|\Gamma(u)\| < \infty, \qquad (6.1.1)$$

where $\|\Gamma(u)\|$ is the square root of the maximum eigenvalues of $\Gamma(u)\Gamma(u)'$.

Then the spectral density matrix of $\{\mathbf{X}(t)\}$ is given by

$$f(\omega) = \frac{1}{2\pi} \sum_{u=-\infty}^{\infty} \Gamma(u)e^{-iu\omega}.$$

Let $l_1 \geq \ldots \geq l_p$ be the eigenvalues of matrix $C(0) = \frac{1}{n} \sum_{t=1}^{n} \mathbf{X}(t)\mathbf{X}(t)'$, and let $\lambda_1 \geq \ldots \geq \lambda_p$ be the eigenvalues of $\Gamma(0)$. We impose a further assumption on them.

Assumption 6.1.2. The eigenvalues λ_j $(j = 1, \ldots, p)$ satisfy

$$\lambda_1 = \cdots = \lambda_{q_1} = \theta_1,$$

$$\lambda_{q_1+1} = \cdots = \lambda_{q_1+q_2} = \theta_2,$$

$$\vdots$$

$$\lambda_{q_1+\cdots+q_{r-1}+1} = \cdots = \lambda_{q_1+\cdots+q_r} = \theta_r,$$

where $\theta_1 > \cdots > \theta_r$ and $q_1 + \cdots + q_r = p$.

Now, let $T_j(l_1,\ldots,l_p)$ $(j = 1,\ldots,k)$ be four time continuously differentiable functions of $l_1,\ldots,l_p$ about $\lambda_1,\ldots,\lambda_p$. Set

$$a_{j\alpha} = \left.\frac{\partial T_j(l_1,\ldots,l_p)}{\partial l_g}\right|_{l=\lambda},$$

$$a_{j\alpha\beta} = \left.\frac{\partial^2 T_j(l_1,\ldots,l_p)}{\partial l_g \partial l_h}\right|_{l=\lambda},$$

$$a_{j\alpha\beta\gamma} = \left.\frac{\partial^3 T_j(l_1,\ldots,l_p)}{\partial l_g \partial l_h \partial l_t}\right|_{l=\lambda},$$

for $g \in J_\alpha$, $h \in J_\beta$, $t \in J_\gamma$, $\boldsymbol{\lambda}' = (\lambda_1,\ldots,\lambda_p)$ and $\mathbf{l}' = (l_1,\ldots,l_p)$, where J_α denotes the set of integers $q_1 + \cdots + q_{\alpha-1} + 1,\ldots,q_1 + \cdots + q_\alpha$ for $\alpha = 1,2,\ldots,r$. We also set down an assumption concerning the functions $T_j(\cdot)$.

Assumption 6.1.3. For each α $(\alpha = 1,\ldots,r)$, the functions $T_j(l_1,\ldots,l_p)$, $(j = 1,\ldots,k)$, are symmetric with respect to $l_{q_1+\cdots+q_{\alpha-1}+1},\ldots,l_{q_1+\cdots+q_\alpha}$, and the derivatives $\partial T_j/\partial l_g$, $\partial^2 T_j/\partial l_g \partial l_h$ and $\partial^3 T_j/\partial l_g \partial l_h \partial l_t$ do not depend on the choices of $g \in J_\alpha$, $h \in J_\beta$ and $t \in J_\gamma$.

In this section we are interested in obtaining the asymptotic joint distribution of

$$L_j = \sqrt{n}\left\{T_j(l_1,\ldots,l_p) - T_j(\lambda_1,\ldots,\lambda_p)\right\}, \quad (j = 1,\ldots,k).$$

Expanding $T_j(l_1,\ldots,l_p)$ as Taylor series, we obtain

$$
\begin{aligned}
L_j &= \sqrt{n}\sum_{\alpha=1}^r a_{j\alpha}\mathrm{tr}\,(W_\alpha - \theta_\alpha I) + \frac{\sqrt{n}}{2}\sum_{\alpha=1}^r\sum_{\beta=1}^r a_{j\alpha\beta}\mathrm{tr}\,(W_\alpha - \theta_\alpha I)\mathrm{tr}\,(W_\beta - \theta_\beta I) \\
&\quad + \frac{\sqrt{n}}{6}\sum_{\alpha=1}^r\sum_{\beta=1}^r\sum_{\gamma=1}^r a_{j\alpha\beta\gamma}\mathrm{tr}\,(W_\alpha - \theta_\alpha I)\mathrm{tr}\,(W_\beta - \theta_\beta I)\mathrm{tr}\,(W_\gamma - \theta_\gamma I) \\
&\quad + \text{higher order terms},
\end{aligned}
$$

where W_α is a random matrix with eigenvalues l_g, $g \in J_\alpha$, and I is an identity matrix. If we assume that W_α is of the form

$$W_\alpha = \theta_\alpha I + \frac{1}{\sqrt{n}}W_\alpha^{(1)} + \frac{1}{n}W_\alpha^{(2)} + \cdots,$$

then L_j can be written (see Fujikoshi(1977), Krishnaiah and Lee(1979))

$$L_j = \sum_{\alpha=1}^{r} a_{j\alpha} \operatorname{tr} W_{\alpha}^{(1)} + \frac{1}{\sqrt{n}} \left\{ \sum_{\alpha=1}^{r} a_{j\alpha} \operatorname{tr} W_{\alpha}^{(2)} + \frac{1}{2} \sum_{\alpha=1}^{r} \sum_{\beta=1}^{r} a_{j\alpha\beta} \operatorname{tr} W_{\alpha}^{(1)} \operatorname{tr} W_{\beta}^{(1)} \right\}$$
$$+ \text{ higher order terms.}$$

We need the following lemma.

Lemma 6.1.1. (Fujikoshi(1977)). *Let T be a square random matrix and $d_1 \geq \ldots \geq d_p$ be the eigenvalues of T. Also, let $\lambda_1 \geq \ldots \geq \lambda_p$ be the corresponding population eigenvalues satisfying Assumption 6.1.2. In addition, we assume that T can be expressed as*

$$T = \Lambda + \epsilon U^{(1)} + \epsilon^2 U^{(2)} + \cdots,$$

where $\Lambda = \operatorname{diag}(\lambda_1, \ldots, \lambda_p)$ and $\epsilon > 0$ is very small. Then the eigenvalues d_j ($j \in J_\alpha$ are the eigenvalues of

$$Z_\alpha = \theta_\alpha I_{q_\alpha} + \epsilon Z_\alpha^{(1)} + \epsilon^2 Z_\alpha^{(2)} + \cdots,$$

where

$$Z_\alpha^{(1)} = U_{\alpha\alpha}^{(1)}, \quad Z_\alpha^{(2)} = U_{\alpha\alpha}^{(2)} + \sum_{\beta \neq \alpha} (\theta_\alpha - \theta_\beta)^{-1} U_{\alpha\beta}^{(1)} U_{\beta\alpha}^{(1)},$$

$$U^{(i)} = \begin{pmatrix} U_{11}^{(i)} & U_{12}^{(i)} & \cdots & U_{1r}^{(i)} \\ U_{21}^{(i)} & U_{22}^{(i)} & \cdots & U_{2r}^{(i)} \\ \vdots & & & \vdots \\ U_{r1}^{(i)} & U_{r2}^{(i)} & \cdots & U_{rr}^{(i)} \end{pmatrix},$$

and $U_{\alpha\beta}^{(i)}$ is of order $q_\alpha \times q_\beta$.

Since there exists an orthogonal matrix B such that $B'\Gamma(0)B = \Lambda$ we put $T = B'C(0)B$ and $U = \sqrt{n}(T - \Lambda)$. Applying Lemma 6.1.1 to U we have

$$L_j = \sum_{\alpha=1}^{r} a_{j\alpha} \operatorname{tr} U_{\alpha\alpha} + \frac{1}{\sqrt{n}} \left\{ \sum_{\alpha=1}^{r} a_{j\alpha} \operatorname{tr} \sum_{\beta \neq \alpha} (\theta_\alpha - \theta_\beta)^{-1} U_{\alpha\beta} U_{\beta\alpha} \right.$$
$$\left. + \frac{1}{2} \sum_{\alpha=1}^{r} \sum_{\beta=1}^{r} a_{j\alpha\beta} \operatorname{tr} U_{\alpha\alpha} \operatorname{tr} U_{\beta\beta} \right\} + \text{ higher order terms.} \tag{6.1.2}$$

We denote the (α, β) components of $C(0)$, $\Gamma(0)$ and $f(\omega)$ by $C(\alpha, \beta)$, $\Gamma(\alpha, \beta)$ and $f_{\alpha\beta}(\omega)$, respectively. To evaluate the asymptotic cumulants of L_j we need the following lemma (e.g., Brillinger(1975), Hosoya and Taniguchi(1982)).

Lemma 6.1.2. *Under Assumptions 6.1.1, the asymptotic covariance between $\sqrt{n}\{C(\alpha_1, \alpha_2) - \Gamma(\alpha_1, \alpha_2)\}$ and $\sqrt{n}\{C(\alpha_3, \alpha_4) - \Gamma(\alpha_3, \alpha_4)\}$ is given by*

$$2\pi \int_{-\pi}^{\pi} \left\{ f_{\alpha_1\alpha_3}(\omega)\overline{f_{\alpha_2\alpha_4}(\omega)} + f_{\alpha_1\alpha_4}(\omega)\overline{f_{\alpha_2\alpha_3}(\omega)} \right\} d\omega.$$

To avoid unnecessary complexity, without loss of generality, we assume

Assumption 6.1.4. The process $\mathbf{Y}(t) = B'\mathbf{X}(t)$, $t = 0, \pm 1, \ldots$ has the diagonal spectral density matrix $\tilde{f}(\omega) = \text{diag}\{\tilde{f}_{11}(\omega), \ldots, \tilde{f}_{pp}(\omega)\}$.

Under Assumptions 6.1.4, Lemma 6.1.2 applied to (6.1.2) shows that

$$
\begin{aligned}
E(L_j) &= \frac{1}{\sqrt{n}} \left\{ \sum_{\alpha=1}^{r} \sum_{\beta \neq \alpha} \sum_{i \in J_\alpha} \sum_{k \in J_\beta} a_{j\alpha}(\theta_\alpha - \theta_\beta)^{-1} \cdot 2\pi \int_{-\pi}^{\pi} \tilde{f}_{ii}(\omega)\tilde{f}_{kk}(\omega)\, d\omega \right. \\
&\quad \left. + \frac{1}{2} \sum_{\alpha=1}^{r} \sum_{i \in J_\alpha} a_{j\alpha\alpha} 4\pi \int_{-\pi}^{\pi} \tilde{f}_{ii}(\omega)^2\, d\omega \right\} + o(n^{-\frac{1}{2}}) \\
&= \frac{c_j}{\sqrt{n}} + o(n^{-\frac{1}{2}}), \quad \text{(say)}.
\end{aligned}
\tag{6.1.3}
$$

$$
\begin{aligned}
\text{cum}\{L_j, L_m\} &= \text{cum}\left\{ \sum_{\alpha=1}^{r} \sum_{k \in J_\alpha} a_{j\alpha} U_{kk}, \sum_{\alpha'=1}^{r} \sum_{k' \in J_{\alpha'}} a_{m\alpha'} U_{k'k'} \right\} + o(n^{-\frac{1}{2}}) \\
&= \sum_{\alpha=1}^{r} \sum_{k \in J_\alpha} a_{j\alpha} a_{m\alpha} 4\pi \int_{-\pi}^{\pi} \tilde{f}_{kk}(\omega)^2\, d\omega + o(n^{-\frac{1}{2}}) \\
&= c_{jm} + o(n^{-\frac{1}{2}}), \quad \text{(say)}.
\end{aligned}
\tag{6.1.4}
$$

Now define

$$Z_n(i,j) = \frac{1}{\sqrt{n}} \left[\sum_{t=1}^{n} \{X_i(t) X_j(t) - \Gamma(i,j)\} \right].$$

Using the fundamental properties of cumulants we have

$$
\begin{aligned}
&\text{cum}\{Z_n(i_1,j_1), Z_n(i_2,j_2), Z_n(i_3,j_3)\} \\
&= \frac{1}{n\sqrt{n}} \sum_{\{(l_1,l_2),(l_3,l_4),(l_5,l_6)\}}^{*} \text{tr}\, \Gamma_{l_1l_2}(n \times n)\, \Gamma_{l_3l_4}(n \times n)\, \Gamma_{l_5l_6}(n \times n),
\end{aligned}
\tag{6.1.5}
$$

where $\sum^{*}$ is the sum over all two-dimensional indecomposable partitions of

$$
\begin{pmatrix}
i_1 & j_1 \\
i_2 & j_2 \\
i_3 & j_3
\end{pmatrix},
$$

and $\Gamma_{l_1l_2}(n \times n)$ is an $n \times n$ matrix with $\text{cum}\{X_{l_1}(t), X_{l_2}(s)\}$ in the tth row and sth column. To evaluate (6.1.5) we need a vector version of Theorem 2.2.1.

Lemma 6.1.3. *Let* $\mathbf{X}(t) = (X_1(t), \ldots, X_p(t))'$ *be a Gaussian stationary process which satisfies Assumption 6.1.1, with the spectral density matrix* $f(\omega) = \{f_{ij}(\omega)\}$ *and zero mean. Then*

$$\frac{1}{n} \mathrm{tr}\, \Gamma_{i_1 j_1}(n \times n)\, \Gamma_{i_2 j_2}(n \times n) \cdots \Gamma_{i_s j_s}(n \times n)$$
$$= (2\pi)^{s-1} \int_{-\pi}^{\pi} f_{i_1 j_1}(\omega) f_{i_2 j_2}(\omega) \cdots f_{i_s j_s}(\omega)\, d\omega + O(n^{-1}). \tag{6.1.6}$$

From Lemma 6.1.3 we see that

$$\mathrm{cum}\{L_j, L_m, L_s\} = \frac{32\pi^2}{\sqrt{n}} \sum_{\alpha=1}^{r} \sum_{k \in J_\alpha} a_{j\alpha} a_{m\alpha} a_{s\alpha} \int_{-\pi}^{\pi} \tilde{f}_{kk}(\omega)^3\, d\omega$$
$$+ \frac{8\pi^2}{\sqrt{n}} \sum_{\alpha=1}^{r} \sum_{\beta=1}^{r} \sum_{k_1 \in J_\alpha} \sum_{k_2 \in J_\beta} \int_{-\pi}^{\pi} \tilde{f}_{k_1 k_1}(\omega)^2\, d\omega \int_{-\pi}^{\pi} \tilde{f}_{k_2 k_2}(\omega)^2\, d\omega$$
$$\times \{a_{j\alpha} a_{m\beta} a_{s\alpha\beta} + a_{j\alpha} a_{m\beta} a_{s\beta\alpha} + a_{j\alpha} a_{s\beta} a_{m\alpha\beta} + a_{j\alpha} a_{s\beta} a_{m\beta\alpha}$$
$$+ a_{s\alpha} a_{m\beta} a_{j\alpha\beta} + a_{s\alpha} a_{m\beta} a_{j\beta\alpha}\} + o(n^{-\frac{1}{2}})$$
$$= \frac{1}{\sqrt{n}} c_{jms} + o(n^{-\frac{1}{2}}), \quad \text{(say)}, \tag{6.1.7}$$

and the Jth-order cumulant satisfies

$$\mathrm{cum}^{(J)}\{L_{j_1}, \ldots, L_{j_J}\} = O(n^{-\frac{J}{2}+1}). \quad \text{for each} \quad J \geq 3.$$

Therefore the next theorem follows from (2.1.15).

Theorem 6.1.1. *Under Assumptions 6.1.1 – 6.1.4,*

$$P(L_1 < y_1, \ldots, L_k < y_k)$$
$$= \int_{-\infty}^{y_1} \cdots \int_{-\infty}^{y_k} N(\mathbf{y} : \Omega) \left[1 + \frac{1}{\sqrt{n}} \sum_{j=1}^{k} c_j H_j(\mathbf{y}) \right.$$
$$\left. + \frac{1}{6\sqrt{n}} \sum_{j=1}^{k} \sum_{m=1}^{k} \sum_{s=1}^{k} c_{jms} H_{jms}(\mathbf{y}) \right] d\mathbf{y} + o(n^{-\frac{1}{2}}), \tag{6.1.8}$$

where $\mathbf{y} = (y_1, \ldots, y_k)'$, $N(\mathbf{y} : \Omega) = (2\pi)^{-k/2} |\Omega|^{-1/2} \exp(-\frac{1}{2}\mathbf{y}'\Omega^{-1}\mathbf{y})$,

$$H_{j_1 \cdots j_s}(\mathbf{y}) = \frac{(-1)^s}{N(\mathbf{y} : \Omega)} \frac{\partial^s}{\partial y_{j_1} \cdots \partial y_{j_s}} N(\mathbf{y} : \Omega),$$

and $\Omega = \{c_{jm}\}$ ($k \times k$-*matrix*).

Corollary 6.1.1. Suppose that the eigenvalues of $\Gamma(0)$ satisfy $\lambda_1 > \cdots > \lambda_p > 0$ and $T_j(\lambda_1, \ldots, \lambda_p) = \lambda_j$ in Theorem 6.1.1. In the special case when the spectral densities are constants such that

$$\tilde{f}_{jj}(\omega) = \frac{\lambda_j}{2\pi} \quad (j = 1, \ldots, p)$$

(i.e., the usual multivariate analysis case), the expansion (6.1.8) becomes

$$P\left\{\sqrt{n}(l_1 - \lambda_1) < y_1, \ldots, \sqrt{n}(l_p - \lambda_p) < y_p\right\}$$
$$= \int_{-\infty}^{y_1} \cdots \int_{-\infty}^{y_p} N(\mathbf{y} : \tilde{\Omega}) \left[1 + \frac{1}{\sqrt{n}} \sum_{j=1}^{p} \tilde{c}_j H_j(\mathbf{y}) \right.$$
$$\left. + \frac{1}{6\sqrt{n}} \sum_{j=1}^{p} \sum_{m=1}^{p} \sum_{s=1}^{p} \tilde{c}_{jms} H_{jms}(\mathbf{y}) \right] d\mathbf{y} + o(n^{-\frac{1}{2}}). \tag{6.1.9}$$

Here

$$\tilde{c}_j = \sum_{\beta \neq j} (\lambda_j - \lambda_\beta)^{-1} \lambda_j \lambda_\beta,$$

$$\tilde{c}_{jms} = 8\lambda_j^3 \delta(j, m)\delta(m, s),$$

and the (j, m)-th element of $\tilde{\Omega}$ is

$$\tilde{c}_{jm} = 2\delta(j, m)\lambda_j^2,$$

where $\delta(j, m)$ is Kronecker's delta. This result agrees with that of Sugiura(1976).

For testing problem $H : \lambda_{p-q} > \lambda_{p-q+1} = \cdots = \lambda_p = \theta$ $(\theta > 0)$ against $A : \lambda_{p-q+1} \geq \cdots \geq \lambda_p > 0$, we consider the criterion

$$L = -n \log \frac{\prod_{i=p-q+1}^{p} l_i}{\left(\frac{1}{q}\sum_{i=p-q+1}^{p} l_i\right)^q}.$$

In the usual multivariate analysis this is known as the likelihood ratio criterion. Then we get the asymptotic expansion of L under the nonnull hypothesis.

Corollary 6.1.2. Let $T_1(l_{m+1}, \ldots, l_p) = \frac{1}{n}L$ in Theorem 6.1.1 and $m = p - q$. Then, under the alternative A, we have

$$P\left[\sqrt{n}\left\{T_1(l_{m+1}, \ldots, l_p) - T_1(\lambda_{m+1}, \ldots, \lambda_p)\right\} \leq y_1\right]$$
$$= \int_{-\infty}^{y_1} N(y_1 : \Omega) \left[1 + \frac{1}{\sqrt{n}}c_1 H_1(y_1) + \frac{1}{6\sqrt{n}}c_{111} H_{111}(y_1)\right] dy_1 + o(n^{-\frac{1}{2}}),$$

where c_1, Ω and c_{111} are defined in (6.1.3), (6.1.4) and (6.1.7), respectively, with

$$a_{1\alpha} = \frac{\sum_{i=m+1}^{p}(\lambda_\alpha - \lambda_i)}{\lambda_\alpha \sum_{i=m+1}^{p} \lambda_i},$$

$$a_{1\alpha\alpha} = \frac{1}{\lambda_\alpha^2} - \frac{p - m}{\left(\sum_{i=m+1}^{p} \lambda_i\right)^2}, \quad \alpha = m+1, \ldots, p,$$

and

$$a_{1\alpha\beta} = -\frac{p - m}{\left(\sum_{i=m+1}^{p} \lambda_i\right)^2}, \quad \alpha \neq \beta, \quad \alpha, \beta = m+1, \ldots, p.$$

6.2. Asymptotic expansions of the distributions of functions of the eigenvalues of canonical correlation matrix in multivariate time series

Suppose that we have two large sets of time series and wish to study the interrelations. If the two sets are very large, we may wish to consider only a few linear combinations of each set. Then we are led to consider the canonical correlation analysis of time series. Hannan(1970) and Brillinger(1969, 1975) discussed the canonical correlation analysis in the frequency domain, and investigated various limiting properties of certain statistics. Also Robinson(1973) gave some first-order asymptotic results for statistics related to the sample canonical correlation matrix in multivariate time series. In this section we shall derive the asymptotic expansions for certain functions of the sample canonical correlation matrix in multivariate time series.

Let

$$\mathbf{X}(t)' = (\mathbf{X}_1(t)', \mathbf{X}_2(t)')$$
$$= (X_1(t), \ldots, X_p(t), X_{p+1}(t), \ldots, X_{p+q}(t)), \quad (p \leq q),$$

be a $(p+q)$-vector-valued Gaussian stationary process with zero mean and covariance matrix $\Gamma(j) = E\{\mathbf{X}(t)\mathbf{X}(t+j)'\}$, which satisfies Assumption 6.1.1. We also assume that $\{\mathbf{X}(t)\}$ has the spectral density matrix $f(\omega) = \{f_{\alpha\beta}(\omega)\}$.

Put

$$C(0) = \frac{1}{n}\sum_{t=1}^{n}\mathbf{X}(t)\mathbf{X}(t)' = \begin{pmatrix} S_{11} & S_{12} \\ S_{21} & S_{22} \end{pmatrix},$$

$$\Gamma(0) = E\{C(0)\} = \begin{pmatrix} M_{11} & M_{12} \\ M_{21} & M_{22} \end{pmatrix},$$

$$Y = \sqrt{n}\{C(0) - \Gamma(0)\} = \begin{pmatrix} Y_{11} & Y_{12} \\ Y_{21} & Y_{22} \end{pmatrix}.$$

Define the $p \times q$ matrix G as

$$G = M_{11}^{-\frac{1}{2}} M_{12} M_{22}^{-\frac{1}{2}}.$$

By the singular value decomposition theorem, there exist two orthogonal matrices Γ_1 and Γ_2 of order $p \times p$ and $q \times q$, respectively, such that

$$G = \Gamma_1 P \Gamma_2',$$

where $P = \{\text{diag}\,(\rho_1, \ldots, \rho_p)|\mathbf{0}\}$, $PP' = \text{diag}\,(\lambda_1, \ldots, \lambda_p)$, and $\lambda_i = \rho_i^2$ $(i = 1, \ldots, p)$. Define

$$\pi_1 = \Gamma_1' M_{11}^{-\frac{1}{2}},$$

$$\pi_2 = \Gamma_2' M_{22}^{-\frac{1}{2}}.$$

Then, using an argument similar to Fang and Krishnaiah(1982b), we can see

$$(\pi_1')^{-1} S_{11}^{-1} S_{12} S_{22}^{-1} S_{21} \pi_1' = PP' + \frac{1}{\sqrt{n}}(PV_{21} - PV_{22}P' + V_{12}P' - V_{11}PP')$$

$$+\frac{1}{n}(PV_{22}V_{22}P' - PV_{22}V_{21} + V_{12}V_{21} - V_{12}V_{22}P'$$

$$-V_{11}PV_{21} + V_{11}PV_{22}P' - V_{11}V_{12}P' + V_{11}V_{11}PP')$$

$$+O_p(n^{-\frac{3}{2}}), \tag{6.2.1}$$

where $V_{ij} = \pi_i Y_{ij} \pi_j'$ $(i, j = 1, 2)$.

Now, without loss of generality we assume that $\{\mathbf{X}(t)\}$ has the covariance matrix

$$\Gamma(0) = \begin{pmatrix} 1 & & 0 & \rho_1 & & 0 & \\ & \ddots & & & \ddots & & 0 \\ 0 & & 1 & 0 & & \rho_p & \\ \rho_1 & & 0 & 1 & & & 0 \\ & \ddots & & & & & \\ 0 & & \rho_p & & & \ddots & \\ & & & & & & \\ & 0 & & 0 & & & 1 \end{pmatrix}, \tag{6.2.2}$$

and the spectral density matrix

$$\tilde{f}(\omega) = \{\tilde{f}_{jk}(\omega)\} = \begin{pmatrix} f_{11}^{(1)}(\omega) & & 0 & f_{12}^{(1)}(\omega) & & 0 & \\ & \ddots & & & \ddots & & 0 \\ 0 & & f_{11}^{(p)}(\omega) & 0 & & f_{12}^{(p)}(\omega) & \\ f_{21}^{(1)}(\omega) & & 0 & f_{22}^{(1)}(\omega) & & & 0 \\ & \ddots & & & & & \\ 0 & & f_{21}^{(p)}(\omega) & & & \ddots & \\ & & & & & & \\ & 0 & & 0 & & & f_{22}^{(q)}(\omega) \end{pmatrix}, \tag{6.2.3}$$

with

$$\rho_j = \int_{-\pi}^{\pi} f_{21}^{(j)}(\omega)\,d\omega = \int_{-\pi}^{\pi} f_{12}^{(j)}(\omega)\,d\omega, \quad j = 1, \dots, p,$$

$$1 = \int_{-\pi}^{\pi} f_{11}^{(j)}(\omega)\,d\omega, j = 1, \dots, p,$$

$$1 = \int_{-\pi}^{\pi} f_{22}^{(j)}(\omega)\,d\omega, j = 1, \dots, q.$$

Thus we may assume that $\pi_1 = I_p$ and $\pi_2 = I_q$. Let $l_1 \geq \dots \geq l_p$ be the eigenvalues of $S_{11}^{-1} S_{12} S_{22}^{-1} S_{21}$, and suppose that the functions $T_j(l_1, \dots, l_p)$, $(j = 1, \dots, k)$, satisfy Assumptions 6.1.3 with the same notations defined as in Section 6.1. We set down

$$L_j = \sqrt{n}\left\{T_j(l_1, \dots, l_p) - T_j(\lambda_1, \dots, \lambda_p)\right\}; j = 1, \dots, k,$$

where $\lambda_1, \dots, \lambda_p$ satisfy Assumptions 6.1.2. Then, we can proceed as in Section 6.1 to get the stochastic expansion;

$$L_j = \sum_{\alpha=1}^{r} a_{j\alpha} \operatorname{tr} W_\alpha^{(1)} + \frac{1}{\sqrt{n}} \left\{ \sum_{\alpha=1}^{r} a_{j\alpha} \operatorname{tr} W_\alpha^{(2)} + \frac{1}{2} \sum_{\alpha=1}^{r} \sum_{\beta=1}^{r} a_{j\alpha\beta} \operatorname{tr} W_\alpha^{(1)} \cdot \operatorname{tr} W_\beta^{(1)} \right\}$$
$$+ \text{ higher order terms,} \tag{6.2.4}$$

with

$$W_\alpha^{(1)} = U_{\alpha\alpha}^{(1)},$$
$$W_\alpha^{(2)} = U_{\alpha\alpha}^{(2)} + \sum_{\beta \neq \alpha} (\theta_\alpha - \theta_\beta)^{-1} U_{\alpha\beta}^{(1)} U_{\beta\alpha}^{(1)},$$

where

$$U^{(1)} = PY_{21} - PY_{22}P' + Y_{12}P' - Y_{11}PP',$$

$$U^{(2)} = PY_{22}Y_{22}P' - PY_{22}Y_{21} + Y_{12}Y_{21} - Y_{12}Y_{22}P',$$
$$-Y_{11}PY_{21} + Y_{11}PY_{22}P' - Y_{11}Y_{12}P' + Y_{11}Y_{11}PP'.$$

Define

$$U(k,m) = \rho_k Y(p+k,m) - \rho_k \rho_m Y(p+k,p+m) + \rho_m Y(k,p+m) - \rho_m^2 Y(k,m),$$
$$V(m,k) = \rho_k Y(k+p,m+p) - Y(k,m+p),$$

where $Y(\alpha, \beta)$ is the (α, β)-th component of the matrix Y. Then, by (6.2.4) we have

$$L_j = \sum_{\alpha=1}^{r} a_{j\alpha} \left\{ \sum_{k \in J_\alpha} U(k,k) \right\} + \frac{1}{\sqrt{n}} \left\{ \sum_{\alpha=1}^{r} a_{j\alpha} \left[\sum_{i=1}^{r+1} \sum_{k \in J_\alpha} \sum_{m \in J_i} V(m,k)^2 \right. \right.$$
$$- \sum_{i=1}^{r} \sum_{k \in J_\alpha} \sum_{m \in J_i} Y(k,m) U(m,k) + \sum_{\beta \neq \alpha} (\theta_\alpha - \theta_\beta)^{-1} \sum_{k \in J_\alpha} \sum_{m \in J_\alpha} U(k,m) U(m,k) \Bigg]$$
$$+ \frac{1}{2} \sum_{\alpha=1}^{r} \sum_{\beta=1}^{r} a_{j\alpha\beta} \sum_{k \in J_\alpha} \sum_{m \in J_\alpha} U(k,k) U(m,m) \Bigg\} + \text{ higher order terms,} \tag{6.2.5}$$

where J_{r+1} denotes the set of integers $p+1, \ldots, q$.
We denote

$$K_{vv}(m,k:m,k) = E\{V(m,k)^2\},$$
$$K_{yu}(k,m:m,k) = E\{Y(k,m)U(m,k)\},$$
$$K_{uu}(k,m:m,k) = E\{U(k,m)U(m,k)\},$$
$$K_{uv}(l,l:m,k) = E\{U(l,l)V(m,k)\},$$
$$K_{uuu}(k:m:s) = \sqrt{n}\,\operatorname{cum}\{U(k,k),U(m,m),U(s,s)\}.$$

It follows from Lemma 6.1.2 that

$$
\begin{aligned}
K_{vv}(m, k : m, k) =\ & 2\pi\rho_k^2 \int_{-\pi}^{\pi} \{1 + \delta(m,k)\} f_{22}^{(k)}(\omega) f_{22}^{(m)}(\omega)\, d\omega \\
& -4\pi\rho_k \int_{-\pi}^{\pi} \left\{ f_{21}^{(k)}(\omega) f_{22}^{(m)}(\omega) + \delta(m,k) f_{22}^{(k)}(\omega) f_{12}^{(m)}(\omega) \right\} d\omega \\
& +2\pi \int_{-\pi}^{\pi} \left\{ f_{11}^{(k)}(\omega) f_{22}^{(m)}(\omega) + \delta(m,k) f_{12}^{(k)}(\omega)^2 \right\} d\omega + O(n^{-1}), \qquad (6.2.6)
\end{aligned}
$$

$$
\begin{aligned}
K_{yu}(k, m : l, l) =\ & \Bigg[4\pi\rho_l \int_{-\pi}^{\pi} f_{11}^{(l)}(\omega) \left\{ f_{12}^{(l)}(\omega) + f_{21}^{(l)}(\omega) \right\} d\omega \\
& -4\pi\rho_l^2 \int_{-\pi}^{\pi} f_{12}^{(l)}(\omega) f_{21}^{(l)}(\omega)\, d\omega \\
& -4\pi\rho_l^2 \int_{-\pi}^{\pi} f_{11}^{(l)}(\omega)^2\, d\omega \Bigg] \delta(k,l)\delta(m,l) + O(n^{-1}), \qquad (6.2.7)
\end{aligned}
$$

$$
\begin{aligned}
K_{uu}(k, m : m, k) =\ & 2\pi\rho_k^2\rho_m^2 \int_{-\pi}^{\pi} \{1 + \delta(m,k)\} \left\{ f_{11}^{(k)}(\omega) f_{11}^{(m)}(\omega) + f_{22}^{(k)}(\omega) f_{22}^{(m)}(\omega) \right\} d\omega \\
& -2\pi\rho_k\rho_m^2 \int_{-\pi}^{\pi} \Big[\left\{ f_{12}^{(k)}(\omega) + \delta(m,k) f_{21}^{(k)}(\omega) \right\} f_{22}^{(m)}(\omega) \\
& \qquad + \{1 + \delta(m,k)\} f_{12}^{(k)}(\omega) f_{22}^{(m)}(\omega) + \left\{ f_{21}^{(k)}(\omega) + \delta(m,k) f_{12}^{(k)}(\omega) \right\} f_{22}^{(m)}(\omega) \Big] d\omega \\
& -2\pi\rho_k^2\rho_m \int_{-\pi}^{\pi} \Big[\{1 + \delta(m,k)\} \left\{ f_{11}^{(k)}(\omega) f_{12}^{(m)}(\omega) + f_{22}^{(k)}(\omega) f_{21}^{(m)}(\omega) \right\} \\
& \qquad + \left\{ f_{12}^{(m)}(\omega) + \delta(m,k) f_{21}^{(m)}(\omega) \right\} f_{22}^{(k)}(\omega) \Big] d\omega \\
& -2\pi\rho_k^3 \int_{-\pi}^{\pi} \{1 + \delta(m,k)\} f_{21}^{(k)}(\omega) f_{11}^{(m)}(\omega)\, d\omega \\
& -2\pi\rho_m^3 \int_{-\pi}^{\pi} \left\{ f_{21}^{(m)}(\omega) + \delta(m,k) f_{12}^{(m)}(\omega) \right\} f_{11}^{(k)}(\omega)\, d\omega \\
& +2\pi\rho_k^2 \int_{-\pi}^{\pi} \left\{ f_{22}^{(k)}(\omega) f_{11}^{(m)}(\omega) + \delta(m,k) f_{21}^{(k)}(\omega) f_{21}^{(m)}(\omega) \right\} d\omega \\
& +2\pi\rho_m^2 \int_{-\pi}^{\pi} \left\{ f_{22}^{(m)}(\omega) f_{11}^{(k)}(\omega) + \delta(m,k) f_{12}^{(k)}(\omega) f_{12}^{(m)}(\omega) \right\} d\omega \\
& +2\pi\rho_k\rho_m \int_{-\pi}^{\pi} \Big\{ f_{21}^{(k)}(\omega) f_{21}^{(m)}(\omega) + \delta(m,k) f_{22}^{(k)}(\omega) f_{11}^{(m)}(\omega) \\
& \qquad + f_{12}^{(k)}(\omega) f_{12}^{(m)}(\omega) + \delta(m,k) f_{11}^{(k)}(\omega) f_{22}^{(m)}(\omega) \Big\} d\omega \\
& +2\pi\rho_k^3\rho_m \int_{-\pi}^{\pi} f_{21}^{(k)}(\omega) f_{12}^{(m)}(\omega)\{1 + \delta(m,k)\}d\omega \\
& +2\pi\rho_k\rho_m^3 \int_{-\pi}^{\pi} f_{12}^{(k)}(\omega) f_{21}^{(m)}(\omega)\{1 + \delta(m,k)\}d\omega + O(n^{-1}), \qquad (6.2.8)
\end{aligned}
$$

$$
K_{uu}(m, k : l, l) = \begin{pmatrix} O(n^{-1}) & \text{for all } m \neq k, l \\ O(n^{-1}) & \text{for all } m = k \neq l, \end{pmatrix}
$$

$$
\begin{aligned}
K_{uv}(l, l : m, k) =\ & 10\pi\rho_l^2 \int_{-\pi}^{\pi} f_{22}^{(l)}(\omega) f_{21}^{(l)}(\omega)\, d\omega \\
& -4\pi\rho_l^3 \int_{-\pi}^{\pi} \left\{ f_{22}^{(l)}(\omega)^2 + f_{12}^{(l)}(\omega) f_{21}^{(l)}(\omega) \right\} d\omega \\
& -4\pi\rho_l \int_{-\pi}^{\pi} \left\{ f_{21}^{(l)}(\omega)^2 + f_{22}^{(l)}(\omega) f_{11}^{(l)}(\omega) \right\} d\omega
\end{aligned}
$$

$$+2\pi\rho_l^2 \int_{-\pi}^{\pi} \left\{ f_{22}^{(l)}(\omega)f_{12}^{(l)}(\omega) + \left(f_{21}^{(l)}(\omega) + f_{12}^{(l)}(\omega) \right) f_{11}^{(l)}(\omega) \right\} d\omega$$
$$+O(n^{-1}) \quad \text{for } l = m = k,$$
$$= O(n^{-1}) \quad \text{otherwise.} \tag{6.2.9}$$

Let Δ be the set of integers k and $p+k$, and let

$$\gamma(j_1, j_2) = \begin{pmatrix} \rho_k & \text{if } |j_1 - j_2| = p, \\ -\rho_k^2 & \text{if } |j_1 - j_2| = 0. \end{pmatrix}$$

Then, using Lemma 6.1.3 we have

$$K_{uuu}(k : m : s) = (2\pi)^2 \sum_{j_1,\ldots,j_6 \in \Delta} \gamma(j_1, j_2)\gamma(j_3, j_4)\gamma(j_5, j_6)$$
$$\times \sum_{\nu}^{*} \int_{-\pi}^{\pi} \tilde{f}_{\nu_1\nu_2}(\omega)\tilde{f}_{\nu_3\nu_4}(\omega)\tilde{f}_{\nu_5\nu_6}(\omega) d\omega + O(n^{-1}) \quad \text{for } k = m = s,$$
$$= 0 \quad \text{otherwise,} \tag{6.2.10}$$

where $\sum_{\nu}^{*}$ is the sum of all two-dimensional indecomposable partitions of $\begin{pmatrix} (j_1, j_2) \\ (j_3, j_4) \\ (j_5, j_6) \end{pmatrix}$. Thus noting (6.2.5) – (6.2.10) we can show that

$$E(L_j) = \frac{1}{\sqrt{n}}c_j^{(1)} + o(n^{-\frac{1}{2}}), \tag{6.2.11}$$

$$\text{cum}\{L_j, L_s\} = c_{js}^{(2)} + o(n^{-\frac{1}{2}}), \tag{6.2.12}$$

$$\text{cum}\{L_j, L_s, L_l\} = \frac{1}{\sqrt{n}}c_{jsl}^{(3)} + o(n^{-\frac{1}{2}}), \tag{6.2.13}$$

$$\text{cum}^{(J)}\{L_{j_1}, \ldots, L_{j_J}\} = O(n^{-\frac{J}{2}+1}), \quad \text{for each } J \geq 3, \tag{6.2.14}$$

where

$$c_j^{(1)} = \sum_{\alpha=1}^{r} a_{j\alpha} \left\{ \sum_{i=1}^{r+1} \sum_{k \in J_\alpha} \sum_{m \in J_i} K_{vv}(m, k : m, k) - \sum_{i=1}^{r} \sum_{k \in J_\alpha} \sum_{m \in J_i} K_{yu}(k, m : m, k) \right.$$
$$\left. + \sum_{\beta \neq \alpha} (\theta_\alpha - \theta_\beta)^{-1} \sum_{k \in J_\alpha} \sum_{m \in J_\beta} K_{uu}(k, m : m, k) \right\}$$
$$+ \frac{1}{2} \sum_{\alpha=1}^{r} a_{j\alpha\alpha} \sum_{k \in J_\alpha} K_{uu}(k, k : k, k), \tag{6.2.15}$$

$$c_{js}^{(2)} = \sum_{\alpha=1}^{r} a_{j\alpha} a_{s\alpha} \sum_{k \in J_\alpha} K_{uu}(k, k : k, k), \tag{6.2.16}$$

$$c_{jsl}^{(3)} = \sum_{\alpha=1}^{r} a_{j\alpha}a_{s\alpha}a_{l\alpha} \sum_{k \in J_\alpha} \{K_{uuu}(k : k : k) + 6K_{uv}(k, k : k, k)^2$$
$$-6K_{uy}(k, k : k, k)K_{uu}(k, k : k, k)\}$$
$$+ \sum_{\alpha=1}^{r} \sum_{\beta=1}^{r} \{a_{j\alpha}a_{s\beta}a_{l\alpha\beta} + a_{j\beta}a_{l\alpha}a_{s\alpha\beta} + a_{l\beta}a_{s\alpha}a_{j\alpha\beta}\}$$
$$\times \sum_{k \in J_\alpha} \sum_{m \in J_\beta} K_{uu}(k, k : k, k)K_{uu}(m, m : m, m). \tag{6.2.17}$$

From (2.1.15) we have

Theorem 6.2.1.

$$P(L_1 < y_1, \ldots, L_k < y_k) = \int_{-\infty}^{y_1} \cdots \int_{-\infty}^{y_k} N(\mathbf{y} : \Omega) \left[1 + \frac{1}{\sqrt{n}} \sum_{j=1}^{k} c_j^{(1)} H_j(\mathbf{y}) \right.$$

$$\left. + \frac{1}{6\sqrt{n}} \sum_{j=1}^{k} \sum_{s=1}^{k} \sum_{l=1}^{k} c_{jsl}^{(3)} H_{jsl}(\mathbf{y}) \right] d\mathbf{y} + o(n^{-\frac{1}{2}}), \qquad (6.2.18)$$

where

$$\Omega = \{c_{js}^{(2)}\} \quad (k \times k\text{-matrix}).$$

Now consider the test of

$$H : \rho_k^2 > \rho_{k+1}^2 = \cdots = \rho_p^2 = 0,$$

with

$$f_{12}^{(k+1)}(\omega) = \cdots = f_{12}^{(p)}(\omega) = 0,$$

and

$$f_{11}^{(k+1)}(\omega) = \cdots = f_{11}^{(p)}(\omega) = f_{22}^{(k+1)}(\omega) = \cdots = f_{22}^{(q)}(\omega) = \frac{1}{2\pi},$$

against

$$A : \rho_{k+1}^2 \geq \cdots \geq \rho_p^2 \geq 0 \quad \text{and} \quad \rho_{k+1}^2 > 0.$$

For this testing problem we use the following statistic

$$Q = -n \log \prod_{j=k+1}^{p} (1 - l_j). \qquad (6.2.19)$$

In the usual multivariate analysis (6.2.19) is known as the likelihood ratio statistic for testing H. Then, under H, it is not difficult to show

$$Q = \sum_{l=k+1}^{p} \sum_{j=k+1}^{q} Y(l, p+j)^2 + \text{ higher order terms}.$$

Thus we have

Proposition 6.2.1. When the null hypothesis H is true, the limiting distribution as $n \to \infty$ of Q is the χ^2-distribution with $(p - k)(q - k)$ degrees of freedom.

Using Theorem 6.2.1 we can get the asymptotic expansion of Q under the non-null hypothesis.

Proposition 6.2.2. Let $T_1(l_{k+1}, \ldots, l_p) = \frac{1}{n}Q$ in Theorem 6.2.1. Then, under A, we have

$$P\left[\sqrt{n}\left\{T_1(l_{k+1}, \ldots, l_p) - T_1(\rho_{k+1}^2, \ldots, \rho_p^2)\right\} < y_1\right]$$

$$= \int_{-\infty}^{y_1} N(y_1; \Omega)\left[1 + \frac{c_1^{(1)}}{\sqrt{n}}H_1(y_1) + \frac{c_{111}^{(3)}}{6\sqrt{n}}H_{111}(y_1)\right] dy_1 + o(n^{-\frac{1}{2}}),$$

where $c_1^{(1)}$, Ω and $c_{111}^{(3)}$ are defined in (6.2.15), (6.2.16) and (6.2.17), with

$$a_{1\alpha} = \frac{1}{1 - \rho_\alpha^2}, \quad a_{1\alpha\alpha} = \frac{1}{(1 - \rho_\alpha^2)^2}, \quad \alpha = k + 1, \ldots, p,$$

and $a_{1\alpha\beta} = 0$, $\alpha \neq \beta$.

The results discussed in this chapter are developed further by Taniguchi and Maekawa(1990). Let $\{X(t)\}$ be a multivariate Gaussian stationary process with the spectral density matrix $f_\theta(\omega)$, where θ is an unknown parameter vector. Using a quasi-maximum likelihood estimator $\hat{\theta}$ of θ they derived the asymptotic expansions of the distributions of functions of $f_{\hat{\theta}}(\omega)$. They also gave the asymptotic expansions for the distributions of functions of the eigenvalues of $f_{\hat{\theta}}(\omega)$.

CHAPTER 7

SOME PRACTICAL EXAMPLES

In this chapter we verify our higher order asymptotic theory for time series analysis by some numerical studies. The results agree with the theory, which means that our higher order asymptotic theory is practically useful.

7.1. The second-order asymptotic bias

The asymptotic bias is one of the most fundamental and practical concept in the asymptotic theory. We have already evaluated the second-order asymptotic bias for various estimators. Here we give numerical studies of some estimators and their bias corrections.

Example 7.1. For the ARMA model discussed in Chapter 2, we evaluated the second-order bias of the maximum likelihood estimator (MLE) and the quasi-maximum likelihood estimator (q-MLE) (see (2.2.36) and (2.2.55)). Consider the following AR(1)-model,

$$X_t = \theta X_{t-1} + \epsilon_t, \quad |\theta| < 1, \tag{7.1}$$

where $\{\epsilon_t\}$ is a sequence of i.i.d. $N(0,1)$ random variables. Let $\hat{\theta}_{ML}$ and $\hat{\theta}_{qML}$ be the MLE and q-MLE of θ based on $(X_1, \ldots, X_n)'$, respectively. Their bias were evaluated up to second order, i.e.,

$$E_\theta(\hat{\theta}_{ML}) = \theta - \frac{2\theta}{n} + o(n^{-1}), \tag{7.2}$$

$$E_\theta(\hat{\theta}_{qML}) = \theta - \frac{3\theta}{n} + o(n^{-1}), \tag{7.3}$$

(see (2.2.64) and (2.2.68)). Thus $\hat{\theta}_{qML}$ has greater bias than $\hat{\theta}_{ML}$. In view of (7.2) and (7.3) if we make the bias-corrected estimators $\hat{\theta}^*_{ML} = \hat{\theta}_{ML} + 2\hat{\theta}_{ML}/n$ and $\hat{\theta}^*_{qML} = \hat{\theta}_{qML} + 3\hat{\theta}_{qML}/n$, then we can expect that they will be better than $\hat{\theta}_{ML}$ and $\hat{\theta}_{qML}$, respectively. In what follows we use the following approximations;

$$\hat{\theta}_{ML} \doteq (1 - n^{-1}) \frac{\sum_{t=2}^n X_t X_{t-1}}{\sum_{t=2}^{n-1} X_t^2}, \tag{7.4}$$

$$\hat{\theta}_{qML} \doteq \frac{\sum_{t=1}^{n-1} X_t X_{t+1}}{\sum_{t=1}^n X_t^2}. \tag{7.5}$$

These approximations are eligible because the right hand sides of (7.4) and (7.5) are known to be asymptotically equivalent to $\hat{\theta}_{ML}$ and $\hat{\theta}_{qML}$ up to third order, respectively (see p.53 and p.32). For $n = 100$ we computed $\hat{\theta}_{ML}$, $\hat{\theta}^*_{ML}$, $\hat{\theta}_{qML}$ and $\hat{\theta}^*_{qML}$ for $\theta = -0.9 \ (0.3) \ 0.9$. Table 7.1 gives the averaged values of $\hat{\theta}_{ML}$, $\hat{\theta}^*_{ML}$, $\hat{\theta}_{qML}$ and $\hat{\theta}^*_{qML}$ by 30 trials simulation for $\theta = -0.9 \ (0.3) \ 0.9$.

Table 7.1:

θ	$\hat{\theta}_{ML}$	$\hat{\theta}_{ML}^{*}$	$\hat{\theta}_{qML}$	$\hat{\theta}_{qML}^{*}$
-0.9	-0.873	-0.891	-0.864	-0.890
-0.6	-0.584	-0.595	-0.577	-0.594
-0.3	-0.291	-0.297	-0.288	-0.297
-0.0	-0.001	-0.001	-0.002	-0.002
0.3	0.288	0.294	0.284	0.293
0.6	0.577	0.589	0.570	0.587
0.9	0.870	0.887	0.859	0.885

From the table we observe that $\hat{\theta}_{qML}$ has greater bias than $\hat{\theta}_{ML}$, and that the bias-corrected estimators $\hat{\theta}_{ML}^{*}$ and $\hat{\theta}_{qML}^{*}$ are better estimators than $\hat{\theta}_{ML}$ and $\hat{\theta}_{qML}$, respectively. Also it may be noted that the values of $\hat{\theta}_{ML}^{*}$ and $\hat{\theta}_{qML}^{*}$ are very near.

Example 7.2. Our higher order asymptotic theory has been developed for a stationary multivariate time series $\{\mathbf{X}(t) = (X_1(t), \ldots, X_p(t))'\}$. Suppose that a stretch $\{\mathbf{X}(1), \ldots, \mathbf{X}(n)\}$ is available. We define

$$\Gamma(0) = E\{\mathbf{X}(t)\mathbf{X}(t)'\},$$

$$C(0) = \frac{1}{n}\sum_{t=1}^{n}\mathbf{X}(t)\mathbf{X}(t)'.$$

Let $l_1 \geq \ldots \geq l_p$ and $\lambda_1 \geq \ldots \geq \lambda_p$ are the eigenvalues of $C(0)$ and $\Gamma(0)$, respectively. In Chapter 6, the formula (6.1.3) gives the second-order bias of a function $T(l_1, \ldots, l_p)$ of $l_1, \ldots, l_p$. Here we consider the following bivariate process;

$$\mathbf{X}(t) = \begin{pmatrix} X_1(t) \\ X_2(t) \end{pmatrix} = \begin{pmatrix} \frac{\sqrt{3}}{2}\left\{\epsilon_t + \sqrt{\lambda-1}\cdot\epsilon_{t-1}\right\} - \frac{1}{2}\eta_t \\ \frac{1}{2}\left\{\epsilon_t + \sqrt{\lambda-1}\cdot\epsilon_{t-1}\right\} + \frac{\sqrt{3}}{2}\eta_t \end{pmatrix},$$

where $\lambda > 1$, and $\{(\epsilon_t, \eta_t)'\}$ is a sequence of i.i.d. $N\left(\begin{pmatrix} 0 \\ 0 \end{pmatrix}, \begin{pmatrix} 1 & 0 \\ 0 & 1 \end{pmatrix}\right)$ vector random variables. Then it is easy to check that the eigenvalues of $\Gamma(0)$ are λ and 1. To estimate λ we use the largest eigenvalue l_1 of $C(0)$. It follows from (6.1.3) that

$$E(l_1) = \lambda + \frac{1}{n}\frac{\lambda}{\lambda-1} + o(n^{-1}). \tag{7.6}$$

In view of (7.6) we propose the bias-corrected estimator $l_1^{*} = l_1 - n^{-1}l_1/(l_1 - 1)$. For $n = 50$ we computed l_1 and l_1^{*} for $\lambda = 1.1$ (0.1) 2.0. Table 7.2 gives the averaged values of l_1 and l_1^{*} by 30 trials simulation for $\lambda = 1.1$ (0.1) 2.0.

Table 7.2:

λ	l_1	l_1^*
1.1	1.244	1.278
1.2	1.318	1.173
1.3	1.397	1.324
1.4	1.480	1.381
1.5	1.567	1.498
1.6	1.657	1.597
1.7	1.749	1.695
1.8	1.843	1.793
1.9	1.938	1.892
2.0	2.034	1.990

Table 7.2 shows that the bias-corrected estimator l_1^* is better than l_1 (except for $\lambda = 1.1$). Therefore it is worth taking the second-order bias-correction into consideration.

7.2. Edgeworth approximations

Here we study the accuracy of Edgeworth type expansions of the maximum likelihood estimator and the quasi-maximum likelihood estimator by numerical simulation.

Example 7.3. Suppose that a stretch $(X_1, \ldots, X_n)'$ of the process (7.1) is available. To estimate θ we use the MLE $\hat{\theta}_{ML}$ and q-MLE $\hat{\theta}_{qML}$ given by (7.4) and (7.5), respectively. Then it is known that the Edgeworth expansions for $\hat{\theta}_{ML}$ and $\hat{\theta}_{qML}$ are given by

$$P_\theta^n \left\{ \frac{\sqrt{n}(\hat{\theta}_{ML} - \theta)}{\sqrt{1 - \theta^2}} \leq x \right\} = \Phi(x) - \phi(x) \left[\frac{-\theta(x^2 + 1)}{\sqrt{n(1 - \theta^2)}} \right] + o(n^{-\frac{1}{2}}), \qquad (7.7)$$

$$P_\theta^n \left\{ \frac{\sqrt{n}(\hat{\theta}_{qML} - \theta)}{\sqrt{1 - \theta^2}} \leq x \right\} = \Phi(x) - \phi(x) \left[\frac{-\theta(x^2 + 2)}{\sqrt{n(1 - \theta^2)}} \right] + o(n^{-\frac{1}{2}}), \qquad (7.8)$$

respectively (see p.25 and p.29), where $\phi(x) = (2\pi)^{-1/2} \exp -\frac{x^2}{2}$ and $\Phi(x) = \int_{-\infty}^{x} \phi(t)\, dt$. For $n = 100$ we computed the probabilities $P_\theta^n\{\sqrt{n}(\hat{\theta}_{ML} - \theta)/\sqrt{1 - \theta^2} \leq x\}$ and $P_\theta^n\{\sqrt{n}(\hat{\theta}_{qML} - \theta)/\sqrt{1 - \theta^2} \leq x\}$ by 1000 trials simulation for $\theta = 0.3, 0.6, 0.9$. In Figures 7.1 – 7.3, we plotted $F(x) = P_\theta^n\{\sqrt{n}(\hat{\theta}_{ML} - \theta)/\sqrt{1 - \theta^2} \leq x\}$ together with the graphs of $Nor(x) = \Phi(x)$ and $Edg(x) = \Phi(x) - \phi(x)[-\theta(x^2 + 1)/\sqrt{n(1 - \theta^2)}]$, for $\theta = 0.3, 0.6, 0.9$, respectively. Similarly, in Figures 7.4 – 7.6, we plotted $F(x) = P_\theta^n\{\sqrt{n}(\hat{\theta}_{qML} - \theta)/\sqrt{1 - \theta^2} \leq x\}$ together with the graphs of $Nor(x) = \Phi(x)$ and $Edg(x) = \Phi(x) - \phi(x)[-\theta(x^2 + 2)/\sqrt{n(1 - \theta^2)}]$, for $\theta = 0.3, 0.6, 0.9$, respectively. The results show that the second-order Edgeworth approximations $Edg(x)$ give better approximations than the usual normal approximations $Nor(x)$. Therefore the higher order approximations in the distribution are also useful in time series analysis.

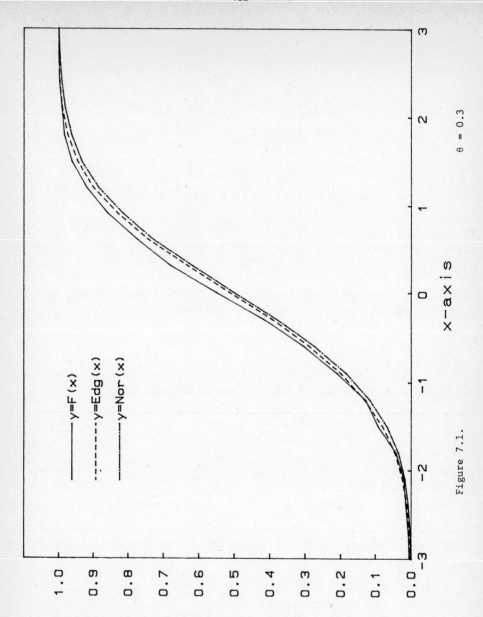

y=F(x)

y=Edg(x)

y=Nor(x)

x-axis

θ = 0.3

Figure 7.1.

133

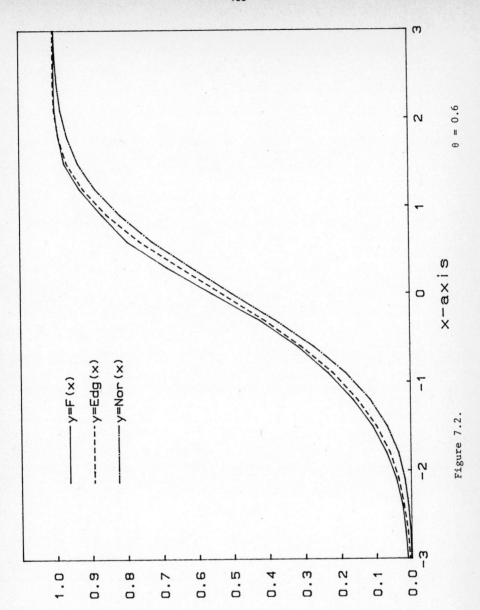

x-axis

Figure 7.2.

θ = 0.6

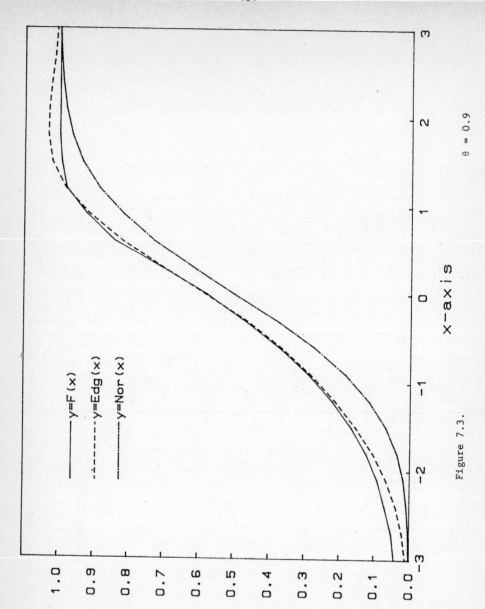

x-axis

θ = 0.9

Figure 7.3.

y=F(x)
y=Edg(x)
y=Nor(x)

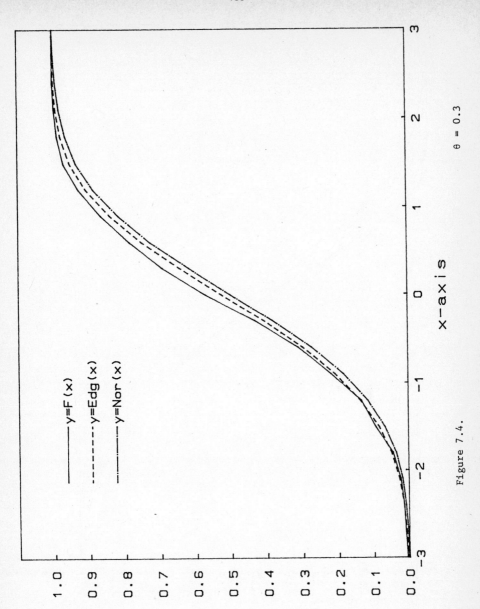

x-axis

y=F(x)

y=Edg(x)

y=Nor(x)

θ = 0.3

Figure 7.4.

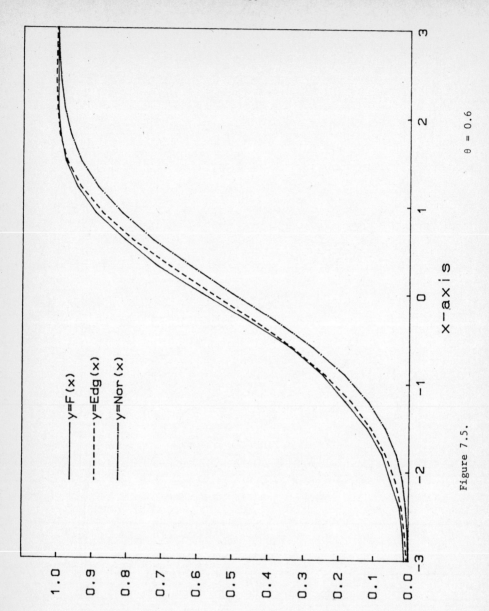

θ = 0.6

Figure 7.5.

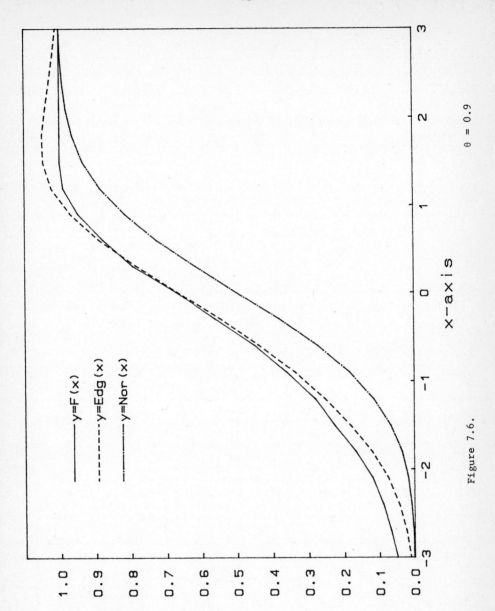

y=F(x)

y=Edg(x)

y=Nor(x)

x-axis

θ = 0.9

Figure 7.6.

7.3. Testing problems

In Chapter 5 we investigated the higher order testing theory for a Gaussian ARMA process. Here we discuss Bartlett's adjustments and the second-order (local) power comparisons by numerical studies.

Example 7.4. Consider the AR(1)-model defined by (7.1). Let LR be the likelihood ratio test for the testing problem $H : \theta = \theta_0$ against $A : \theta \neq \theta_0$. From Theorem 5.1.1 the asymptotic expansion of LR under H is given by

$$P_{\theta_0}^n[LR \leq x] = P[\chi_1^2 \leq x] + \frac{1}{n}\left[P\{\chi_1^2 \leq x\} - P\{\chi_3^2 \leq x\}\right] + o(n^{-1}). \tag{7.9}$$

Since Bartlett's adjustment factor for LR is 2 (see p.109), the adjusted LR test is given by $LR^\star = (1 + \frac{2}{n})LR$. Thus its asymptotic expansion is

$$P_{\theta_0}^n[LR^\star \leq x] = P[\chi_1^2 \leq x] + o(n^{-1}), \tag{7.10}$$

(i.e., the n^{-1}-order terms vanish). From (7.9) and (7.10) we can expect that χ_1^2-approximation of $LR^\star$ is better than that of LR. Now we verify this numerically. Let x_α be the level α point of χ_1^2 (i.e., $P[\chi_1^2 > x_\alpha] = \alpha$). For $n = 100$ we computed the following probabilities;

$$P(0.05) = P_{\theta_0}^n[LR > x_{0.05}], \qquad P(0.025) = P_{\theta_0}^n[LR > x_{0.025}],$$
$$P(0.01) = P_{\theta_0}^n[LR > x_{0.01}], \qquad BP(0.05) = P_{\theta_0}^n[LR^\star > x_{0.05}],$$
$$BP(0.025) = P_{\theta_0}^n[LR^\star > x_{0.025}], \quad BP(0.01) = P_{\theta_0}^n[LR^\star > x_{0.01}],$$

by 1000 trials simulation for $\theta_0 = -0.9\ (0.3)\ 0.9$. These values are given by the table below.

Table 7.3:

θ_0	$P(0.05)$	$P(0.025)$	$P(0.01)$	$BP(0.05)$	$BP(0.025)$	$BP(0.01)$
-0.9	0.051	0.021	0.006	0.051	0.024	0.008
-0.6	0.041	0.027	0.011	0.044	0.028	0.012
-0.3	0.037	0.020	0.009	0.042	0.021	0.009
0.0	0.039	0.017	0.008	0.042	0.018	0.009
0.3	0.037	0.021	0.007	0.040	0.021	0.007
0.6	0.043	0.022	0.013	0.048	0.022	0.014
0.9	0.057	0.029	0.008	0.060	0.032	0.011

Table 7.3 shows that Bartlett's adjustment is effective to attain the level probability. Since Bartlett's adjustment is a sort of third-order correction in the sense of higher order asymptotic theory, the study above shows that this is also useful.

In Section 5.2 we compared the second-order (local) powers of various tests for a Gaussian ARMA process. Here we discuss the power properties of the likelihood ratio test and Rao's test.

Example 7.5. For the AR(1)-model (7.1), let LR and R be the likelihood ratio test and Rao's test for the testing problem $H : \theta = \theta_0$ against $A : \theta \neq \theta_0$, respectively. Under the sequence of alternatives $A_n : \theta = \theta_0 + \epsilon/\sqrt{n}$, $(\epsilon > 0)$, the second-order difference of power between LR and R is given by

$$P^n_{\theta_0+\epsilon/\sqrt{n}}[R > x] - P^n_{\theta_0+\epsilon/\sqrt{n}}[LR > x]$$
$$= \frac{1}{\sqrt{n}} \frac{6\theta_0}{(1-\theta_0^2)^2} \left\{ \frac{\epsilon^3}{3} p_7(x; \delta) + \frac{\epsilon}{I(\theta_0)} p_5(x; \delta) \right\} + o(n^{-\frac{1}{2}}), \qquad (7.11)$$

where $p_j(x; \delta)$ is the probability density function of $\chi_j^2(\delta)$, $\delta^2 = I(\theta_0)\epsilon^2/2$ and $I(\theta_0) = (1 - \theta_0^2)^{-1}$ (see Example 5.2.4). Thus LR is more powerful than R if $\theta_0 < 0$ and vice versa. Recalling Examples 5.2.2 – 5.2.4, we can see that none of the famous tests is uniformly superior. However, if we modify a test S belonging to a class $\mathbf{S}_A$ defined in Section 5.2 (p.109) to be second-order asymptotically unbiased, then the second-order (local) powers of all the modified test $S^\star$ are identical (see Theorem 5.2.3). In this case LR is second-order asymptotically unbiased, but R is not so. From the modification procedure given in Theorem 5.2.3, it follows that

$$R^\star = \left\{ 1 + (\hat{\theta}_{ML} - \theta_0) \frac{-2\theta_0}{1 - \theta_0^2} \right\} R$$

is second-order unbiased. Suppose that a stretch $(X_1, \ldots, X_{200})'$ is observed from (7.1) under the sequence of alternative $A'_n : \theta = \theta_0 + \epsilon/\sqrt{I(\theta_0)n}$. For each $\epsilon = 2.5, 3.0, 3.2$, we computed the probabilities;

$$PR = P^n_{\theta_0+\epsilon/\sqrt{I(\theta_0)n}}[R > x_{0.05}],$$

$$PR^\star = P^n_{\theta_0+\epsilon/\sqrt{I(\theta_0)n}}[R^\star > x_{0.05}],$$

$$PLR = P^n_{\theta_0+\epsilon/\sqrt{I(\theta_0)n}}[LR > x_{0.05}],$$

by 1000 trials simulation for $\theta_0 = -0.9\ (0.2) - 0.3$. Tables 7.4 – 7.6 give the values of them for $\epsilon = 2.5, 3.0, 3.2$, respectively.

Table 7.4:

$\epsilon = 2.5$

θ_0	PR	$PR^\star$	PLR
-0.9	0.189	0.582	0.562
-0.7	0.475	0.624	0.614
-0.5	0.562	0.633	0.632
-0.3	0.626	0.658	0.661

Table 7.5:

$$\epsilon = 3.0$$

θ_0	PR	$PR^\star$	PLR
-0.9	0.295	0.717	0.698
-0.7	0.641	0.781	0.774
-0.5	0.745	0.811	0.814
-0.3	0.788	0.817	0.815

Table 7.6:

$$\epsilon = 3.2$$

θ_0	PR	$PR^\star$	PLR
-0.9	0.345	0.778	0.754
-0.7	0.706	0.825	0.825
-0.5	0.805	0.853	0.854
-0.3	0.838	0.863	0.866

Tables 7.4 – 7.6 show that LR is more powerful than R if $\theta_0 < 0$, and that the modified test $R^\star$ is as good as LR. Thus the results confirm Theorem 5.2.3, and show that the modification procedure is very effective.

Finally, as a conclusion we can say that our higher order asymptotic theory is practically useful.

TABLE 2.4

VALUES OF L(ρ,x), R(ρ,x) and M(ρ,x)

FOR n = 300, ρ = -0.9(0.3)0.9

x =	-2.00	-1.50	-1.00	-0.50	0.00	0.50	1.00	1.50	2.00
L(-0.9, x)	.0217	.0524	.0822	.0907	.0767	.0818	.0885	.0575	.0512
R(-0.9, x)	.0051	.0176	.0278	.0403	.0327	.0288	.0257	.0012	.0040
M(-0.9, x)	.0195	.0438	.0504	.0431	.0338	.0450	.0647	.0413	.0406
L(-0.6, x)	.0109	.0250	.0252	.0279	.0241	.0290	.0239	.0201	.0156
R(-0.6, x)	.0005	.0070	.0034	.0065	.0078	.0084	.0053	.0021	.0022
M(-0.6, x)	.0091	.0188	.0126	.0103	.0079	.0140	.0155	.0149	.0124
L(-0.3, x)	.0049	.0106	.0092	.0167	.0083	.0190	.0109	.0097	.0096
R(-0.3, x)	.0001	.0040	.0014	.0055	.0018	.0122	.0007	.0027	.0054
M(-0.3, x)	.0041	.0086	.0046	.0077	.0018	.0148	.0069	.0083	.0086
L(0.0, x)	.0004	.0001	.0066	.0052	.0137	.0101	.0037	.0049	.0030
R(0.0, x)	.0006	.0011	.0064	.0052	.0137	.0099	.0041	.0055	.0038
M(0.0, x)	.0006	.0009	.0060	.0058	.0137	.0093	.0047	.0053	.0030
L(0.3, x)	.0076	.0075	.0073	.0066	.0145	.0041	.0070	.0066	.0075
R(0.3, x)	.0044	.0004	.0016	.0011	.0072	.0016	.0021	.0003	.0039
M(0.3, x)	.0068	.0061	.0033	.0008	.0072	.0008	.0026	.0038	.0067
L(0.6, x)	.0156	.0173	.0283	.0220	.0180	.0207	.0260	.0206	.0139
R(0.6, x)	.0003	.0044	.0081	.0042	.0004	.0011	.0038	.0058	.0027
M(0.6, x)	.0116	.0107	.0203	.0090	.0002	.0037	.0142	.0158	.0119
L(0.9, x)	.0444	.0665	.0839	.0724	.0856	.0859	.0742	.0530	.0205
R(0.9, x)	.0030	.0043	.0143	.0194	.0411	.0327	.0248	.0206	.0069
M(0.9, x)	.0348	.0497	.0549	.0378	.0428	.0383	.0442	.0418	.0201

(This table is due to Taniguchi, Krishnaiah and Chao(1989))

Table 2.5.1 $D_1(\alpha,\rho,n)$ n = 30

α \ ρ	-0.875	-0.750	-0.625	-0.500	-0.375	-0.250	-0.125	0.000	0.125	0.250	0.375	0.500	0.625	0.750	0.875
-0.875	0.000	0.001	0.005	0.011	0.020	0.032	0.046	0.062	0.081	0.102	0.126	0.153	0.182	0.214	0.248
-0.750	0.001	0.000	0.001	0.003	0.007	0.012	0.019	0.028	0.038	0.050	0.063	0.078	0.094	0.112	0.131
-0.625	0.003	0.001	0.000	0.001	0.003	0.006	0.010	0.016	0.023	0.032	0.041	0.052	0.065	0.078	0.093
-0.500	0.006	0.002	0.001	0.000	0.001	0.002	0.006	0.010	0.015	0.022	0.030	0.040	0.050	0.062	0.075
-0.375	0.010	0.006	0.003	0.001	0.000	0.001	0.003	0.006	0.010	0.016	0.023	0.031	0.041	0.052	0.064
-0.250	0.018	0.011	0.006	0.003	0.001	0.000	0.001	0.003	0.006	0.011	0.018	0.026	0.035	0.046	0.058
-0.125	0.030	0.021	0.013	0.008	0.003	0.001	0.000	0.001	0.003	0.008	0.013	0.021	0.030	0.041	0.054
0.000	0.051	0.037	0.026	0.017	0.009	0.004	0.001	0.000	0.001	0.004	0.009	0.017	0.026	0.037	0.051
0.125	0.088	0.068	0.050	0.035	0.022	0.012	0.006	0.001	0.000	0.001	0.006	0.012	0.022	0.035	0.051
0.250	0.160	0.126	0.097	0.071	0.049	0.032	0.018	0.006	0.001	0.000	0.002	0.008	0.018	0.032	0.050
0.375	0.310	0.251	0.199	0.152	0.112	0.078	0.050	0.028	0.012	0.003	0.000	0.003	0.012	0.028	0.049
0.500	0.672	0.556	0.450	0.356	0.272	0.200	0.139	0.089	0.050	0.022	0.006	0.000	0.006	0.022	0.050
0.625	1.750	1.471	1.216	0.985	0.778	0.596	0.438	0.304	0.194	0.109	0.049	0.012	0.000	0.012	0.049
0.750	6.438	5.485	4.610	3.810	3.086	2.438	1.867	1.371	0.952	0.610	0.343	0.152	0.038	0.000	0.038
0.875	55.751	48.071	40.960	34.418	28.444	23.040	18.204	13.938	10.240	7.111	4.551	2.560	1.138	0.284	0.000

Table 2.5.2 $D_2(\alpha,\rho,\lambda,n)$ n = 30 $\lambda = \pi/6$

α \ ρ	-0.875	-0.750	-0.625	-0.500	-0.375	-0.250	-0.125	0.000	0.125	0.250	0.375	0.500	0.625	0.750	0.875
-0.875	0.000	0.005	0.020	0.045	0.081	0.126	0.182	0.248	0.323	0.409	0.505	0.611	0.728	0.854	0.991
-0.750	0.003	0.000	0.003	0.012	0.028	0.050	0.078	0.112	0.152	0.198	0.251	0.310	0.375	0.447	0.524
-0.625	0.010	0.003	0.000	0.003	0.010	0.023	0.041	0.064	0.093	0.126	0.165	0.208	0.257	0.311	0.370
-0.500	0.022	0.010	0.002	0.000	0.002	0.010	0.022	0.039	0.061	0.088	0.119	0.156	0.197	0.243	0.294
-0.375	0.040	0.022	0.010	0.002	0.000	0.002	0.010	0.022	0.040	0.062	0.090	0.122	0.160	0.202	0.249
-0.250	0.068	0.043	0.024	0.011	0.003	0.000	0.003	0.011	0.024	0.043	0.068	0.097	0.133	0.173	0.219
-0.125	0.111	0.077	0.049	0.028	0.012	0.003	0.000	0.003	0.012	0.028	0.049	0.077	0.111	0.151	0.197
0.000	0.179	0.131	0.091	0.058	0.033	0.015	0.004	0.000	0.004	0.015	0.033	0.058	0.091	0.131	0.179
0.125	0.285	0.218	0.160	0.111	0.071	0.040	0.018	0.004	0.000	0.004	0.018	0.040	0.071	0.111	0.160
0.250	0.451	0.357	0.273	0.201	0.139	0.089	0.050	0.022	0.006	0.000	0.006	0.022	0.050	0.089	0.139
0.375	0.701	0.568	0.448	0.343	0.252	0.175	0.112	0.063	0.028	0.007	0.000	0.007	0.028	0.063	0.112
0.500	1.038	0.858	0.695	0.549	0.420	0.309	0.215	0.137	0.077	0.034	0.009	0.000	0.009	0.034	0.077
0.625	1.393	1.170	0.967	0.783	0.619	0.474	0.348	0.242	0.155	0.087	0.039	0.010	0.000	0.010	0.039
0.750	1.580	1.346	1.131	0.935	0.757	0.598	0.458	0.337	0.234	0.150	0.084	0.037	0.009	0.000	0.009
0.875	1.637	1.411	1.203	1.010	0.835	0.676	0.534	0.409	0.301	0.209	0.134	0.075	0.033	0.008	0.000

Table 2.5.3 $D_2(\alpha,\rho,\lambda,n)$ n = 30   λ = π/2

α \ ρ	-0.875	-0.750	-0.625	-0.500	-0.375	-0.250	-0.125	0.000	0.125	0.250	0.375	0.500	0.625	0.750	0.875
-0.875	0.000	0.005	0.020	0.045	0.081	0.126	0.181	0.247	0.322	0.408	0.503	0.609	0.725	0.851	0.987
-0.750	0.003	0.000	0.003	0.012	0.027	0.049	0.076	0.110	0.149	0.195	0.247	0.305	0.369	0.439	0.515
-0.625	0.010	0.002	0.000	0.002	0.010	0.022	0.039	0.061	0.089	0.120	0.157	0.199	0.246	0.297	0.354
-0.500	0.020	0.009	0.002	0.000	0.002	0.009	0.020	0.036	0.056	0.080	0.109	0.142	0.180	0.222	0.269
-0.375	0.034	0.019	0.009	0.002	0.000	0.002	0.009	0.019	0.034	0.053	0.077	0.104	0.136	0.172	0.213
-0.250	0.052	0.033	0.019	0.008	0.002	0.000	0.002	0.008	0.019	0.033	0.052	0.075	0.102	0.134	0.169
-0.125	0.075	0.052	0.033	0.019	0.008	0.002	0.000	0.002	0.008	0.019	0.033	0.052	0.075	0.102	0.133
0.000	0.102	0.075	0.052	0.033	0.019	0.008	0.002	0.000	0.002	0.008	0.019	0.033	0.052	0.075	0.102
0.125	0.133	0.102	0.075	0.052	0.033	0.019	0.008	0.002	0.000	0.002	0.008	0.019	0.033	0.052	0.075
0.250	0.169	0.134	0.102	0.075	0.052	0.033	0.019	0.008	0.002	0.000	0.002	0.008	0.019	0.033	0.052
0.375	0.213	0.172	0.136	0.104	0.077	0.053	0.034	0.019	0.009	0.002	0.000	0.002	0.009	0.019	0.034
0.500	0.269	0.222	0.180	0.142	0.109	0.080	0.056	0.036	0.020	0.009	0.002	0.000	0.002	0.009	0.020
0.625	0.354	0.297	0.246	0.199	0.157	0.120	0.089	0.061	0.039	0.022	0.010	0.002	0.000	0.003	0.010
0.750	0.515	0.439	0.369	0.305	0.247	0.195	0.149	0.110	0.076	0.049	0.027	0.012	0.003	0.000	0.003
0.875	0.987	0.851	0.725	0.609	0.503	0.408	0.322	0.247	0.181	0.126	0.081	0.045	0.020	0.005	0.000

Table 2.5.4 $D_2(\alpha,\rho,\lambda,n)$ n = 30   λ = 5π/6

α \ ρ	-0.875	-0.750	-0.625	-0.500	-0.375	-0.250	-0.125	0.000	0.125	0.250	0.375	0.500	0.625	0.750	0.875
-0.875	0.000	0.008	0.033	0.075	0.134	0.209	0.301	0.409	0.534	0.676	0.835	1.010	1.203	1.411	1.637
-0.750	0.000	0.000	0.009	0.037	0.084	0.150	0.234	0.337	0.458	0.598	0.757	0.935	1.131	1.346	1.580
-0.625	0.039	0.010	0.000	0.010	0.039	0.087	0.155	0.242	0.348	0.474	0.619	0.783	0.967	1.170	1.393
-0.500	0.077	0.034	0.009	0.000	0.009	0.034	0.077	0.137	0.215	0.309	0.420	0.549	0.695	0.858	1.038
-0.375	0.112	0.063	0.028	0.007	0.000	0.007	0.028	0.063	0.112	0.175	0.252	0.343	0.448	0.568	0.701
-0.250	0.139	0.089	0.050	0.022	0.006	0.000	0.006	0.022	0.050	0.089	0.139	0.201	0.273	0.357	0.451
-0.125	0.160	0.111	0.071	0.040	0.018	0.004	0.000	0.004	0.018	0.040	0.071	0.111	0.160	0.218	0.285
0.000	0.179	0.131	0.091	0.058	0.033	0.015	0.004	0.000	0.004	0.015	0.033	0.058	0.091	0.131	0.179
0.125	0.197	0.151	0.111	0.077	0.049	0.028	0.012	0.003	0.000	0.003	0.012	0.028	0.049	0.077	0.111
0.250	0.219	0.173	0.133	0.097	0.068	0.043	0.024	0.011	0.003	0.000	0.003	0.011	0.024	0.043	0.068
0.375	0.249	0.202	0.160	0.122	0.090	0.062	0.040	0.022	0.010	0.002	0.000	0.002	0.010	0.022	0.040
0.500	0.294	0.243	0.197	0.156	0.119	0.088	0.061	0.039	0.022	0.010	0.002	0.000	0.002	0.010	0.022
0.625	0.370	0.311	0.257	0.208	0.165	0.126	0.093	0.064	0.041	0.023	0.010	0.003	0.000	0.003	0.010
0.750	0.524	0.447	0.375	0.310	0.251	0.198	0.152	0.112	0.078	0.050	0.028	0.012	0.003	0.000	0.003
0.875	0.991	0.854	0.728	0.611	0.505	0.409	0.323	0.248	0.182	0.126	0.081	0.045	0.020	0.005	0.000

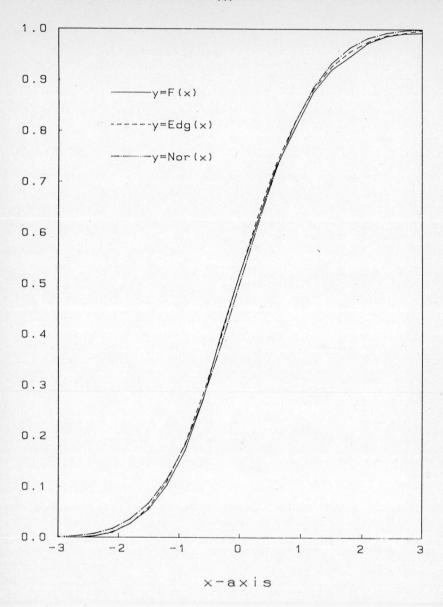

Figure 4.2.1   α = 0.0

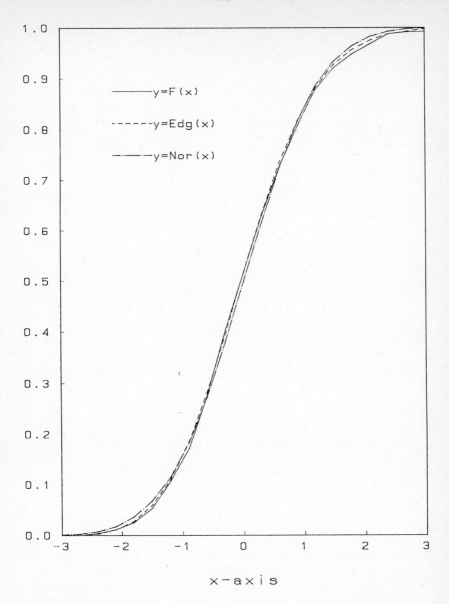

Figure 4.2.2   α = 0.3

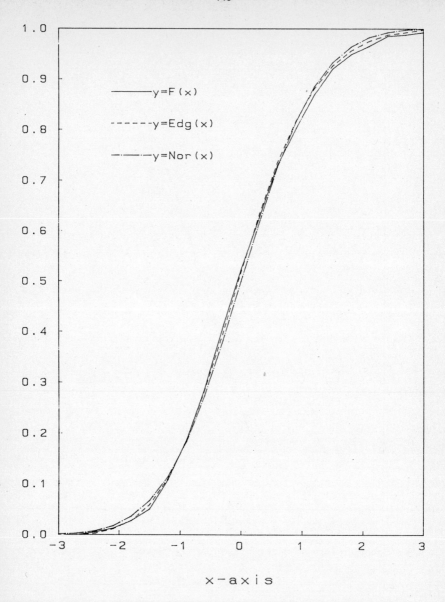

Figure 4.2.3   α = 0.6

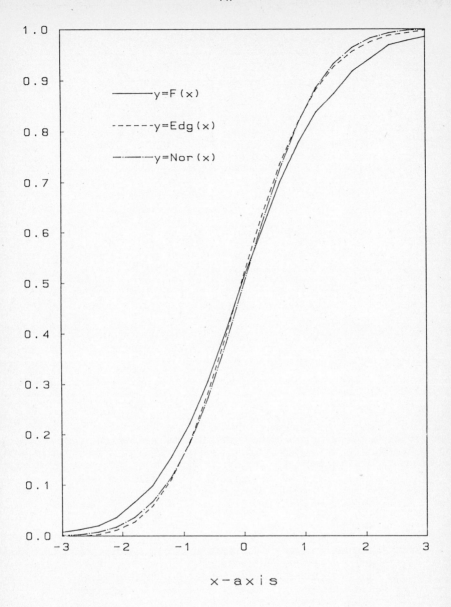

Figure 4.2.4   α = 0.9

Table 4.2.1

α	I_1		I_2	
-0.9	[-0.9719	-0.8454]	[-0.9712	-0.8461]
-0.8	[-0.8940	-0.7246]	[-0.8937	-0.7249]
-0.7	[-0.8087	-0.6076]	[-0.8090	-0.6073]
-0.6	[-0.7187	-0.4933]	[-0.7195	-0.4924]
-0.5	[-0.6249	-0.3805]	[-0.6261	-0.3793]
-0.4	[-0.5279	-0.2689]	[-0.5294	-0.2674]
-0.3	[-0.4285	-0.1587]	[-0.4303	-0.1569]
-0.2	[-0.3273	-0.0502]	[-0.3293	-0.0482]
-0.1	[-0.2251	0.0562]	[-0.2272	0.0583]
0.0	[-0.1221	0.1603]	[-0.1243	0.1625]
0.1	[-0.0186	0.2619]	[-0.0209	0.2641]
0.2	[0.0852	0.3610]	[0.0830	0.3632]
0.3	[0.1897	0.4576]	[0.1876	0.4597]
0.4	[0.2947	0.5517]	[0.2928	0.5536]
0.5	[0.4006	0.6429]	[0.3990	0.6445]
0.6	[0.5072	0.7310]	[0.5061	0.7321]
0.7	[0.6144	0.8147]	[0.6137	0.8154]
0.8	[0.7220	0.8927]	[0.7218	0.8928]
0.9	[0.8328	0.9643]	[0.8334	0.9637]

Table 4.2.2

α	I_3		I_4	
-0.9	[-0.9654	-0.8483]	[-0.9663	-0.8474]
-0.8	[-0.8900	-0.7269]	[-0.8906	-0.7263]
-0.7	[-0.8055	-0.6093]	[-0.8056	-0.6092]
-0.6	[-0.7160	-0.4947]	[-0.7157	-0.4950]
-0.5	[-0.6226	-0.3818]	[-0.6219	-0.3825]
-0.4	[-0.5258	-0.2702]	[-0.5249	-0.2711]
-0.3	[-0.4266	-0.1599]	[-0.4255	-0.1610]
-0.2	[-0.3256	-0.0514]	[-0.3243	-0.0527]
-0.1	[-0.2235	0.0550]	[-0.2220	0.0535]
0.0	[-0.1206	0.1590]	[-0.1191	0.1574]
0.1	[-0.0173	0.2605]	[-0.0157	0.2589]
0.2	[0.0866	0.3594]	[0.0881	0.3579]
0.3	[0.1910	0.4558]	[0.1924	0.4544]
0.4	[0.2960	0.5497]	[0.2973	0.5484]
0.5	[0.4020	0.6407]	[0.4030	0.6396]
0.6	[0.5088	0.7284]	[0.5095	0.7276]
0.7	[0.6162	0.8116]	[0.6165	0.8113]
0.8	[0.7242	0.8888]	[0.7240	0.8891]
0.9	[0.8360	0.9585]	[0.8351	0.9594]

REFERENCES

Akahira, M. (1975). A note on the second-order asymptotic efficiency of estimators in an autoregressive process. *Rep. Univ. Electro-Commun.*, *26*, 143–149.

Akahira, M. (1979). On the second-order asymptotic optimality of estimators in an autoregressive process. *Rep. Univ. Electro-Commun.*, *29*, 213–218.

Akahira, M. and Takeuchi, K. (1981). *Asymptotic Efficiency of Statistical Estimators ; Concepts and Higher Order Asymptotic Efficiency*. Springer Lecture Notes in Statistics, Vol. 7. Springer-Verlag.

Amari, S. (1982). Differential geometry of curved exponential families — curvatures and information loss. *Ann. Statist.*, *10*, 357–385.

Amari, S. (1985). *Differential - Geometrical Methods in Statistics*. Springer Lecture Notes in Statistics, Vol. 28. Springer-Verlag.

Anderson, T.W. (1971). *The Statistical Analysis of Time Series*. Wiley, New York.

Beran, R. (1977). Minimum Hellinger distance estimates for parametric models. *Ann. Statist.*, *5*, 445–463.

Bhattacharya, R.N. and Rao, R.R. (1976). *Normal Approximation and Asymptotic Expansions*. Wiley, New York.

Bhattacharya, R.N. and Ghosh, J.K. (1978). On the validity of the formal Edgeworth expansion. *Ann. Statist.*, *6*, 434–451.

Bloomfield, P. (1973). An exponential model for the spectrum of a scalar time series. *Biometrika*, *60*, 217–226.

Brillinger, D.R. (1969a). Asymptotic properties of spectral estimates of second order. *Biometrika*, *56*, 375–390.

Brillinger, D.R. (1969b). The canonical analysis of stationary time series. In *Multivariate Analysis — II* (P.R. Krishnaiah, Ed.). Academic Press, New York.

Brillinger, D.R. (1975). *Time Series ; Data Analysis and Theory*. New York : Holt.

Chibisov, D.M. (1972). An asymptotic expansion for the distribution of a statistic admitting an asymptotic expansion. *Theor. Prob. Appl.*, *17*, 620–630.

Cox, D.R. (1980). Local ancillarity. *Biometrika*, *67*, 279–286.

Does, R.J.M.M. (1982). Berry-Esseen theorems for simple linear rank statistics under the null-hypothesis. *Ann. Probab., 10*, 982–991.

Dunsmuir, W. (1979). A central limit theorem for parameter estimation in stationary vector time series and its application to models for a signal observed with noise. *Ann. Statist., 7*, 490–506.

Dunsmuir, W. and **Hannan, E.J.** (1976). Vector linear time series models. *Adv. in Appl. Probab., 8*, 339–364.

Durbin, J. (1980a). Approximations for densities of sufficient estimators. *Biometrika, 67*, 311–333.

Durbin, J. (1980b). The approximate distribution of partial serial correlation coefficients calculated from residuals from regression on Fourier series. *Biometrika, 67*, 335–349.

Dzhaparidze, K.O. (1974). A new method for estimating spectral parameters of a stationary regular time series. *Theor. Prob. Appl., 19*, 122–132.

Efron, B. (1975). Defining the curvature of a statistical problem (with applications to second order efficiency). *Ann. Statist., 3*, 1189-1242.

Efron, B. and **Hinkley, D.V.** (1978). Assessing the accuracy of the maximum likelihood estimator : Observed versus expected Fisher information. *Biometrika, 65*, 457–487.

Eguchi, S. (1983). Second order efficiency of minimum contrast estimators in a curved exponential family. *Ann. Statist., 11*, 793–803.

Erickson, R.V. (1974). L_1 bound for asymptotic normality of m-dependent sums using Stein's technique. *Ann. Probab., 2*, 522-529.

Fang, C. and **Krishnaiah, P.R.** (1982a). Asymptotic joint distributions of the elements of sample covariance matrix. *Statistics and Probability : Essays in Honor of C.R. Rao.* North-Holland Publishing Company, 249–262.

Fang, C. and **Krishnaiah, P.R.** (1982b). Asymptotic distributions of functions of the eigenvalues of some random matrices for nonnormal populations. *J. Multivariate Anal., 12*, 39–63.

Fujikoshi, Y. (1977). Asymptotic expansions for the distributions of some multivariate tests. In *Multivariate Analysis — IV* (P.R. Krishnaiah, Ed.). North-Holland, Amsterdam.

Fujikoshi, Y. and **Ochi, Y.** (1984). Asymptotic properties of the maximum likelihood estimate in the first-order autoregressive process. *Ann. Inst. Statist. Math., 36*, 119–128.

Galbraith, R.F. and **Galbraith, J.I.** (1974). On the inverses of some patterned matrices arising in the theory of stationary time series. *J. Appl. Prob., 11*, 63–71.

Ghosh, J.K. and Subramanyam, K. (1974). Second order efficiency of maximum likelihood estimators. *Sankhya Ser. A36*, 325–358.

Götze, F. and Hipp, C. (1983). Asymptotic expansions for sums of weakly dependent random vectors. *Z. Wahrsch. Verw. Gebiete, 64*, 211–239.

Grenander, U. and Szegö, G. (1958). *Toeplitz Forms and Their Applications*. Berkeley, Univ. of California Press.

Hannan, E.J. (1970). *Multiple Time Series*. Wiley, New York.

Hannan, E.J. (1973). The asymptotic theory of linear time series models. *J. Appl. Prob., 10*, 130–145.

Hayakawa, T. (1975). The likelihood ratio criterion for a composite hypothesis under a local alternative. *Biometrika, 62*, 451–460.

Hayakawa, T. (1977). The likelihood ratio criterion and the asymptotic expansion of its distribution. *Ann. Inst. Statist. Math., 29*, 359–378.

Hayakawa, T. and Puri, M.L. (1985). Asymptotic expansions of the distributions of some test statistics. *Ann. Inst. Statist. Math., 37*, 95–108.

Helmers, R. (1981). A Berry-Esseen theorem for linear combinations of order statistics. *Ann. Probab., 9*, 342–347.

Hille, E. (1959). *Analytic Function Theory*, Vol. I. Ginn, Lexington, Massachusetts.

Hosoya, Y. (1974). Estimation problems on stationary time-series models. *Ph. D. thesis*, Yale Univ..

Hosoya, Y. (1979). High-order efficiency in the estimation of linear processes. *Ann. Statist., 7*, 516–530.

Hosoya, Y. (1980). Conditionality and maximum-likelihood estimation. *Recent Developments in Statistical Inference and Data Analysis* ; K. Matusita, editor, 117–126. North-Holland, Amsterdam.

Hosoya, Y. and Taniguchi M. (1982). A central limit theorem for stationary processes and the parameter estimation of linear processes. *Ann. Statist., 10*, 132–153.

Konishi, S. (1978). An approximation to the distribution of sample correlation coefficient. *Biometrika, 65*, 654–656.

Konishi, S. (1981). Normalizing transformations of some statistics in multivariate analysis. *Biometrika, 68*, 647–651.

Krishnaiah, P.R. (1976). Some recent developments on complex multivariate distributions. *J. Multivariate Anal.*, *6*, 1–30.

Krishnaiah, P.R. and **Lee, J.C.** (1979). On the asymptotic joint distributions of certain functions of the eigenvalues of four random matrices. *J. Multivariate Anal.*, *9*, 248–258.

Kumon, M. and **Amari, S.** (1983). Geometrical theory of higher-order asymptotics of test, interval estimator and conditional inference. *Proc. R. Soc. Lond. A387*, 429–458.

LeCam, L. (1956). On the asymptotic theory of estimation and testing hypothesis. *Proceedings of the Third Berkeley Symposium on Mathematical Statistics and Probability*, University of California Press. Vol. 1, 129–156.

Maekawa, K. (1985). *Asymptotic Theory for Regression Analysis.* Research Books. Hiroshima University (in Japanese).

Magnus, J.R. and **Neudecker, H.** (1979). The commutation matrix ; Some properties and applications. *Ann. Statist.*, *7*, 381–394.

Myint Swe and **Taniguchi, M.** (1990). Higher-order asymptotic properties of a weighted estimator for Gaussian ARMA processes. To appear in *J. Time Ser. Anal.*.

Nishio, A. (1981). A comparison of estimators for time series models. *Surikagaku* (Mathematical Science) No. 219, 41–46 (in Japanese).

Ochi, Y. (1983). Asymptotic expansions for the distribution of an estimator in the first-order autoregressive process. *J. Time Ser. Anal.*, *4*, 57–67.

Peers, H.W. (1971). Likelihood ratio and associated test criteria. *Biometrika*, *58*, 577–587.

Pfanzagl, J. (1972). Further results on asymptotic normality I. *Metrika*, *18*, 174–198.

Pfanzagl, J. (1973). Asymptotic expansions related to minimum contrast estimators. *Ann. Statist.*, *1*, 993–1026.

Pfanzagl, J. and **Wefelmeyer, W.** (1978). A third-order optimum property of the maximum likelihood estimator. *J. Multivariate Anal.*, *8*, 1–29.

Phillips, P.C.B. (1977). Approximations to some finite sample distributions associated with a first-order stochastic difference equation. *Econometrica*, *45*, 463–485.

Phillips, P.C.B. (1978). Edgeworth and saddle point approximations in the first-order noncircular autoregression. *Biometrika*, *65*, 91–98.

Phillips, P.C.B. (1979). Normalizing transformations and expansions for functions of statistics. *Research Notes*. Birmingham Univ..

Priestley, M.B., Rao, T. and Tong, H. (1973). Identification of the structure of multivariate stochastic systems. In *Multivariate Analysis — III* (P.R. Krishnaiah, Ed.). Academic Press, New York.

Rao, C.R. (1962). Efficient estimates and optimum inference procedures in large samples. *J. Roy. Statist. Soc. Ser. B, 24*, 46–72.

Robinson, P.M. (1973). Generalized canonical analysis for time series. *J. Multivariate Anal., 3*, 141–160.

Rothenberg, T.J. (1984). Approximate normality of generalized least squares estimates. *Econometrica, 52*, 811–825.

Shaman, P. (1976). Approximations for stationary covariance matrices and their inverses with application to ARMA models. *Ann. Statist., 4*, 292–301.

Skovgaard, I.M. (1985). A second order investigation of asymptotic ancillarity. *Ann. Statist., 13*, 534–551.

Sugiura, N. (1976). Asymptotic expansions of the distributions of the latent roots and the latent vector of the Wishart and multivariate F matrices. *J. Multivariate Anal., 6*, 500–525.

Suzuki, T. (1978). Asymptotic sufficiency up to higher orders and its applications to statistical tests and estimates. *Osaka J. Math., 15*, 575–588.

Takeuchi, K. (1974). *Tokei-teki Suitei no Zenkinriron* (Asymptotic Theory of Statistical Estimation). Kyoiku-Shuppan (in Japanese).

Takeuchi, K. (1981). Structures of the asymptotic best estimation theory. *Surikagaku* (Mathematical Science) No. 219, 5–16 (in Japanese).

Tanaka, K. (1982). Chi-square approximations to the distributions of the Wald, likelihood ratio and Lagrange multiplier test statistics in time series regression. *Tech. Rep. No. 82*, Kanazawa Univ..

Tanaka, K. (1983). Asymptotic expansions associated with the AR(1) model with unknown mean. *Econometrica, 51*, 1221–1231.

Tanaka, K. (1984). An asymptotic expansion associated with the maximum likelihood estimators in ARMA models. *J. Roy. Statist. Soc. Ser. B, 46*, 58–67.

Taniguchi M. (1979). On estimation of parameters of Gaussian stationary processes. *J. Appl. Prob., 16*, 575–591.

Taniguchi M. (1981). An estimation procedure of parameters of a certain spectral density model. *J. Roy. Statist. Soc. Ser. B, 43*, 34–40.

Taniguchi M. (1983). On the second order asymptotic efficiency of estimators of Gaussian ARMA processes. *Ann. Statist.*, *11*, 157–169.

Taniguchi M. (1984). Validity of Edgeworth expansions for statistics of time series. *J. Time Ser. Anal.*, *5*, 37–51.

Taniguchi M. (1985a). Third order efficiency of the maximum likelihood estimator in Gaussian autoregressive moving average processes. In *Statistical Theory and Data Analysis* (K. Matusita, Ed.). 725–743. North-Holland, Amsterdam.

Taniguchi M. (1985b). An asymptotic expansion for the distribution of the likelihood ratio criterion for a Gaussian autoregressive moving average process under a local alternative. *Econometric Theory*, *1*, 73–84.

Taniguchi M. (1986a). Third order asymptotic properties of maximum likelihood estimators for Gaussian ARMA processes. *J. Multivariate Anal.*, *18*, 1–31.

Taniguchi M. (1986b). Berry-Esseen theorems for quadratic forms of Gaussian stationary processes. *Prob. Th. Rel. Fields*, *72*, 185–194.

Taniguchi M. (1987a). Minimum contrast estimation for spectral densities of stationary processes. *J. Roy. Statist. Soc. Ser. B*, *49*, 315–325.

Taniguchi M. (1987b). Third order asymptotic properties of BLUE and LSE for a regression model with ARMA residual. *J. Time Ser. Anal.*, *8*, 111–114.

Taniguchi M. (1987c). Validity of Edgeworth expansions of minimum contrast estimators for Gaussian ARMA processes. *J. Multivariate Anal.*, *21*, 1–28.

Taniguchi M. (1988a). A Berry-Esseen theorem for the maximum likelihood estimator in Gaussian ARMA processes. In *Statistical Theory and Data Analysis* (K. Matusita, Ed.). 535–549.

Taniguchi M. (1988b). Asymptotic expansions of the distributions of some test statistics for Gaussian ARMA processes. *J. Multivariate Anal.*, *27*, 494–511.

Taniguchi M. and Krishnaiah, P.R. (1987). Asymptotic distributions of functions of the eigenvalues of sample covariance matrix and canonical correlation matrix in multivariate time series. *J. Multivariate Anal.*, *22*, 156–176.

Taniguchi M. and Taniguchi, R. (1988). Asymptotic ancillarity in time series analysis. *J. Japan Statist. Soc.*, *18*, 107–121.

Taniguchi M., Krishnaiah, P.R. and Chao, R. (1989). Normalizing transformations of some statistics of Gaussian ARMA processes. *Ann. Inst. Stat. Math.*, *41*, 187–197.

Taniguchi M. and Maekawa, K. (1990). Asymptotic expansions of the distributions of statistics related to the spectral density matrix in multivariate time series and their applications. *Econometric Theory, 6*, 75–96.

Toyooka, Y. (1982). Second-order expansion of mean squared error matrix of generalized least squares estimator with estimated parameters. *Biometrika, 69*, 269–273.

Toyooka, Y. (1985). Second-order risk comparison of SLSE with GLSE and MLE in a regression with serial correlation. *J. Multivariate Anal., 17*, 107–126.

Toyooka, Y. (1986). Second-order risk structure GLSE and MLE in a regression with a linear process. *Ann. Statist., 14*, 1214–1225.

Wakaki, H. (1986). Comparison of powers of a class of tests for covariance matrices. *Tech. Rep. No. 183*, Hiroshima Statistical Research Group. Hiroshima Univ..

Walker, A.M. (1964). Asymptotic properties of least-square estimates of parameters of the spectrum of a stationary nondeterministic time series. *J. Austral. Math. Soc., 4*, 363–384.

Whittle, P. (1951). *Hypothesis Testing in Time Series Analysis.* Uppsala.

Whittle, P. (1952). Some result in time series analysis. *Skandivanisk Aktuarietidskrift., 35*, 48–60.

Whittle, P. (1962). Gaussian estimation in stationary time series. *Bull. Inst. Internat. Statist., 39-2*, 105–129.

Author Index

Subject Index